The Economic Geography of Megaregions

Keith S. Goldfeld, Editor

Essays and commentary sponsored by the Policy Research
Institute for the Region at the Woodrow Wilson School of
Public and International Affairs, Princeton University, and the
Regional Plan Association.

The Policy Research Institute for the Region was established by
Princeton University and the Woodrow Wilson School of Public
and International Affairs to bring the resources of the University
community to bear on solving the increasingly interdependent
public policy challenges facing New Jersey, metropolitan New York,
and southeastern Pennsylvania.

The Policy Research Institute for the Region
Woodrow Wilson School of Public and International Affairs
Princeton University
Robertson Hall
Princeton, NJ 08544

Cover design by Leslie Goldman
Cover photo courtesy of Liam Gumley, Space Science and
    Engineering Center, University of Wisconsin-Madison
Printed by PrintMedia Communications, Anaheim, CA
Produced by the Office of Communications, Princeton University

The views expressed in this publication are those of the authors
and are not necessarily the views of Princeton University or its
Policy Research Institute for the Region.

Contents

Publications from the Policy Research Institute for the Region

Preface

Anyone from Princeton going north towards New York and Boston or south towards Washington, D.C., could easily get the impression that the Northeast Corridor is a single, highly interconnected economic region. Can this area and others like it really be classified as a new kind of economic entity? Is it legitimate or useful to give them a new name, to call them megaregions or something else that tries to convey the new reality? Can we usefully apply to them the methods of analysis customarily used for states or nations? Can we plan for jobs, housing, or transportation, on a megaregion level? And how does the notion of megaregions in the United States change our concept of the work of government agencies or other key institutions?

On February 9, 2007, the Policy Research Institute for the Region and the Regional Plan Association came together at Princeton University to host a discussion on these questions. The day's discussion, titled "The Economic Geography of Megaregions," provided a forum for experts in urban planning, housing, and economics to think more broadly about the challenges and opportunities created by these new developments.

The Policy Research Institute has tried to serve as a vehicle for breaking down the barriers that divide academic disciplines and political jurisdictions so that public policy issues can be addressed fully and creatively. This roundtable discussion on megaregions was just the kind of conversation we had envisioned. We hope that this volume will encourage others to continue the conversation, so we can move forward not merely as observers but as players shaping the megaregions around us.

Nathan B. Scovronick
Acting Director
Policy Research Institute for the Region, Princeton University

Introduction

Petra Todorovich
Regional Plan Association

What is a megaregion? Is it simply a "fact of nature," as Saskia Sassen posits: the result of adjacent metropolitan areas growing together by sprawled development at the metropolitan fringe? Or are there existing and potential economic interactions and agglomerations within megaregions that have new and important policy implications?

How does thinking at the megaregional scale advance our understanding of the economic performance and competitive advantages of different geographic regions of the United States? Does this new scale present a platform for policymaking or regional cooperation that improves upon the state and local levels of government that dominate land-use planning today? What policy areas are most suited for regional or megaregion-scale coordination?

These are some of the questions that guided "The Economic Geography of Megaregions," the February 9 forum hosted by the Regional Plan Association (RPA) and the Policy Research Institute for the Region at the Woodrow Wilson School of Public and International Affairs of Princeton University. The forum convened civic, business, and public

policy leaders to discuss the economic interactions that help define emergent megaregions in the United States, and the potential for public policies to encourage coordination and governance at this new scale.

The discussion papers in this volume, written by Edward Glaeser of Harvard University and Saskia Sassen of the University of Chicago, advance our understanding of the nature of existing and emergent megaregions in the U.S., and propose an analytical framework for understanding the economic effects of greater integration within megaregions.

This scholarly work backfills the meetings and coordination that have already begun between adjacent metropolitan regions in places like the Northeast, Southern California, the Midwest, and other megaregions across the United States, where planners and policymakers have come together to address common challenges that are too big to deal with at the local or regional level, or require cross-jurisdictional coordination among multiple states and regions.

The identification of emerging megaregions nationally and their implications for America's future growth patterns was spurred by a 2004 graduate studio course at the University of Pennsylvania School of Design co-taught by Robert Yaro, Armando Carbonell, and Jonathan Barnett, called "Plan for America."[1] Students used the 2002 Complete Economic and Data Source from Woods & Poole Economics, Inc, which included population and job projections through the year 2025 by county, and projected the numbers to the year 2050, based on a straight line trend. They estimated that the U.S. population would grow by more than 40 percent to 430 million by the year 2050, and identified eight to 10 "SuperCities" across the nation, where more than half the population and economic growth would take place.[2] Subsequent analysis by Regional Plan Association found that over 70 percent of the projected population and employment growth would occur in 10 of these places nationally.[3] The term SuperCities was replaced by "Megaregions," which more accurately reflected the combined urban, suburban, and rural nature of these places.

Continued research, outreach, and planning around megaregions is being coordinated and supported by RPA's America 2050 initiative, an effort to develop a national framework for growth and development in the 21st century, with an emphasis on investments and policy coordination at the megaregion scale. RPA is working with partners in the Northeast—the Lincoln Institute of Land Policy in Cambridge, the Penn Institute for Urban Research in Philadelphia, and the Policy Research Institute for the Region in Princeton, to name a few—to advance an agenda for mobility, sustainability, and economic competitiveness in the Northeast. Nationally, we have convened the National Committee for America 2050, with representation in each of the 10 megaregions, to develop a national framework for America's growth with broad buy-in across the country.

FROM MEGALOPOLIS TO MEGAREGION: DEFINITIONS AND ORIGINS IN THE NORTHEAST

The Regional Plan Association, an independent planning and research organization for the New York–New Jersey–Connecticut region, was founded in the 1920s to create the first regional plan for New York and its environs. RPA was among the first institutions to recognize the importance of the metropolitan region—an urban form made possible by limited access highways and the dispersed settlement patterns they created. Metropolitan regions came to define the American landscape in the 20th century, and grew thanks not only to highways but investments in commuter rail, public transit, and urban amenities that were made in the pre-war years. RPA is now at the forefront of exploring a new urban form, the "megaregion," which has emerged in part due to the continued spread of development at the edges of metropolitan areas, as well as the increasing interactions and shrinking geographies afforded by Internet technology, high-speed travel, and the declining cost of goods movement.

The prescient French geographer Jean Gottman coined the term, "Megalopolis," in his 1961 book of that name, which examined the contiguous urbanized areas stretching roughly from Portland, Maine, to Richmond, Virginia. RPA's Second Regional Plan, released in multiple volumes in the 1960s, also studied the megalopolis in one of its volumes, *The Region's*

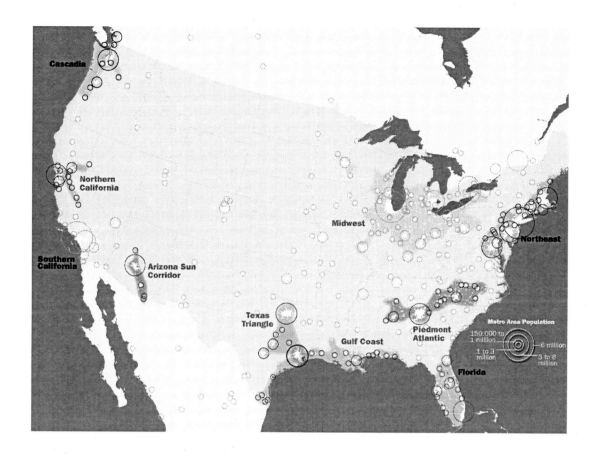

Growth in 1967 and in an edition of *Regional Plan News* in 1969 that discussed development issues and strategies for the "Atlantic Urban Seaboard." RPA's 1969 report documented land use patterns, commuting patterns, and population growth in the Northeast, posing two important questions: (1) Should the region continue to grow at its "most likely rate" or should it encourage diversion of population growth to other parts of the nation? (2) Should policymakers reinforce the region's historic, "nucleated" structure, or should they foster a "spread city," a term for development that was between suburban and rural?

In answering both questions, RPA's research director Boris Pushkarev made a case for why both population growth and greater density in nucleated centers were good for the Northeast. First, he argued that population growth fuels city growth. Specifically, he argued that as societies grow in population size, they require more complex organization of space and bigger "brains" in the form of cities. He foresaw that growth in the megalopolis should contribute to growth in center cities, which in the late 1960s were suffering from increasing rates of poverty, racial segregation, and an outflow of middle- and upper-income residents. He also correctly observed the importance of cities to the knowledge economy, as "machines for communication," which could be measured by their efficiency in providing opportunities for face-to-face contact within some limit of time and cost.

Many of these arguments apply to the emergent megaregions across the United States that RPA observes today, with less dense and complex structures than the Northeast, and more options for the direction of their growth. Some of the most built-out and environmentally vulnerable megaregions, such as Southern California and South Florida, might elect to redirect population growth to other parts of the country (and may be doing so indirectly through local land use regulations, as discussed in Glaeser's paper). However, most other megaregions could benefit from the urban efficiencies and communications advantages of more nucleated urban forms—stronger urban hubs with improved connections between them.

One notable feature of RPA's 1969 report was the observation that while the Northeast megaregion had begun to grow together in land development, its adjacent metropolitan areas still functioned autonomously from one another. In other words, megalopolis remained a "fact of nature": an observation of spatial relationships in aerial photographs. The Northeast metropolitan areas shared historical and cultural trends, similar land development patterns and immigration trends; but as of 1969, these metro areas did not yet exhibit strong evidence of interactions, according to RPA.

That is starting to change. Today, RPA's renewed focus on megaregions is prompted by observations of greater interactions among adjacent metropolitan regions within megaregions—in the Northeast and nationally. These interactions—enabled by overlapping commuting patterns, longer distance flex-time commuting, increased high-speed business travel, rising foreign trade, goods movement, "just-in-time" logistics, and advanced communications technology—put greater pressure on the nation's infrastructure systems, environmental resources, and open space. As we look forward to increased population growth in the United States over the next half-century, these pressures are even more pertinent to the governance structures that guide land development and investments in critical infrastructure.

EXPLORING MEGAREGION BENEFITS: THE NORTHEAST AND BEYOND

To better understand the dynamics and potential of this new, emerging geography, RPA is now pursuing a research and outreach agenda in the Northeast megaregion to explore the existing connections among the five major metro regions that comprise the Northeast—Boston, New York, Philadelphia, Baltimore, and Washington, D.C. We are interested in the potential benefits of fostering greater interactions among these places, through investments in transportation infrastructure, coordination of land use and landscape preservation policies, and actions to reduce carbon emissions and promote greater energy efficiency.

One of the key competitive advantages of the megaregion is the existence of Amtrak's Northeast Corridor, the most intensively used rail corridor in the nation. With approximately

700,000 rides a day (on both commuter and intercity services), the Northeast Corridor boasts the highest passenger use of any rail corridor in the country. The location of its stations in the major and second-tier cities of the Northeast, connected to public transit networks, reinforces the primacy of urban centers and promotes compact development and vibrant downtowns. And as the threat of climate change takes on greater urgency in coming years, the energy efficiency of this mode of travel will grow in importance.

The major cities connected by the corridor—Boston, New York, Philadelphia, Baltimore, Washington, D.C.—are the primary places in the Northeast where face-to-face interaction takes place. Despite the communications enabled by the Internet, the tacit knowledge transmitted by face-to-face interaction is essential to fostering intellectual capital, innovation, collaboration, and trust-building in the global economy. Furthermore, for global firms,

the Northeast is among less than a handful of places in the U.S. that provides access to millions of highly trained workers in specialized fields. As Sassen discusses in the following pages, the headquarter functions of global firms require increasingly specialized labor as their operations grow more and more complex. With the proper investments in intercity and high-speed rail, these workers can travel greater distances with more comfort and greater energy efficiency than anyplace else in the country.

To build a case for increased investment in the Northeast corridor and coordination among the Northeast states, RPA has documented various measures of economic integration in the Northeast megaregion. While the megaregion remains a less integrated scale of geography than the metropolitan region, which is connected by daily commuting patterns, evidence exists of interactions among the component regions of the Northeast. Business travel on

FIGURE I

Employment change index 1990–2004, Northeast regions

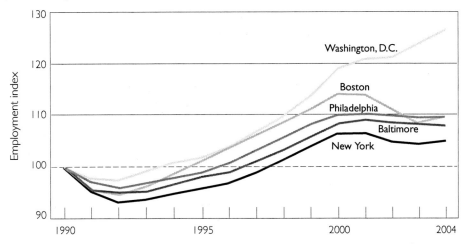

FIGURE 2

Most specialized economic sectors in the Northeast metropolitan regions, as measured by location quotient. (RPA 2006)

Region	Most specialized sector	Second-most specialized
Baltimore	Education and health services	Professional and business services
Boston	Education and health services	Information
New York	Financial activities	Information
Philadelphia	Education and health services	Financial activities
Washington, D.C.	Professional and business services	Information

Amtrak and between airports provide some evidence. There are approximately 26 flights per weekday from Boston to Washington, D.C.; 33 from D.C. to New York; and 33 from New York to Boston, indicating a high level of business travel between the major cities of the Northeast.

Another measure of economic integration is the synchronization of employment cycles over time among five major regions of the Northeast. As shown in Figure 1, employment growth in the five major regional economies follows similar cyclical patterns from 1990 to 2004, which the exception of Washington, D.C., which exhibited more rapid growth since America went to war in 2002. Not surprisingly, the Northeast economies are specialized in many of the same industries, as measured by location quotient.[4] Figure 2 demonstrates the most specialized industries of the five major metropolitan regions of the Northeast. As indicated, all of the regions are specialized in service-based industries, with education and health services, information services, and professional and business services being the most specialized sectors across the board.

The close alignment of these industries suggests the potential for professional interactions among the managerial, legal, training, sales, and creative functions of industries in different regions, which is supported by the travel data. And the complementarities of different industries to each other, i.e., government and defense in Washington, D.C., finance in New York, education and health care in Boston, Baltimore, and Philadelphia, could foster division of labor at the megaregional scale.

To build a case for this theory, and for investments in megaregion-scale infrastructure investments, such as high-speed rail, we asked economist Edward Glaeser of Harvard University and sociologist Saskia Sassen of the University of Chicago to present the following discussion papers at our forum on February 9. The February 9 forum is part of a year-long research program on megaregions as part of the America 2050 initiative. The resulting papers from this program can be found at www.America2050.org.

In the pages that follow, Glaeser presents a comparative study of megaregions in the United States, looking at productivity, housing prices, commuting times, and growth rates. His analysis demonstrates vast variations among the megaregions as measured by these indicators, suggesting to us the value of developing targeted strategies for economic development, education policy, and transportation in different megaregions in the country. His comparative analysis is followed by a discussion of the best scale at which to coordinate a number of public policy issues—education, economic development, housing—concluding that housing affordability is best approached at the regional scale, and potentially the megaregional scale.

Sassen presents an analytical framework for exploring the agglomeration economies of megaregions, pointing to areas for further exploration and analysis. She hypothesizes that megaregions may be sufficiently large and diverse to recover globally outsourced jobs within the megaregion. Firms may find travel, training, communications, and infrastructure costs associated with maintaining overseas operations to be too burdensome over time and choose to repatriate these functions. In the case of the Northeast, they may choose to repatriate these outsourced functions to low-cost areas, such as western New York, central Massachusetts, or rural Pennsylvania, providing local economic benefits to industrial or rural communities that have been struggling to rebuild their economies after the demise of their manufacturing base. But to promote energy and land use efficiencies, these location decisions should be coordinated with transportation investments and land use regulations and incentives.

Recalling RPA's research in the 1960s on cities as the "brains" of increasing complex regions, Sassen's research on global cities has demonstrated that as global firms outsource their functions overseas they rely on increasingly complex headquarters functions with greater demand for highly specialized services. This in turn fuels demand for highly paid workers and high productivity, benefiting the global cities where the headquarters locate. Her paper calls for greater analysis to understand whether these benefits can be extended to the scale of the megaregion. As the paper's title demands, are there greater benefits to megaregions than the simple effect of urban economies of scale?

Glaeser's and Sassen's papers—and the discussion that took place on February 9—only begin to tap the surface of the changing economic, land use, and transportation challenges that will confront the United States as we face rapid population and economic growth in the 21st century. The focus on megaregions stems from need to match the right scale of planning and governance to each challenge that arises. We hope that by focusing on the unique challenges posed by emerging megaregions, we can move toward more effective and sustainable land use policies, as well as tap into the potential economic benefits that a greater mobility of workers, goods, and ideas within megaregions can provide.

Sources

Boris Pushkarev, "The Atlantic Urban Seaboard: Development Issues and Strategies," *Regional Plan News,* September 1969, No 90, New York: Regional Plan Association.

Regional Plan Association, "America 2050: A Prospectus," September 2006, New York.

Jean Gottman, *Megalopolis: The Urbanized Northeastern Seaboard of the United States.* The M.I.T. Press, Cambridge: November 1961.

Notes

1. Robert Yaro is president of the Regional Plan Association and professor-in-practice at the University of Pennsylvania. Armando Carbonell is chair of the Planning and Urban Forum at the Lincoln Institute of Land Policy and a visiting professor at the University of Pennsylvania. Jonathan Barnett is a principal of Wallace Roberts Todd and professor-in-practice at the University of Pennsylvania.

2. The Lincoln Institute of Land Policy, Regional Plan Association, University of Pennsylvania School of Design, "Toward an American Spatial Development Perspective: A Policy Roundtable on the Federal Role in Metropolitan Development," September 2004.

3. Regional Plan Association, "America 2050: A Prospectus," New York: September 2006.

4. Location quotient is calculated as the ratio of an industry's share of the market in a local area, i.e., the metropolitan region, compared to the industry's share in the national economy.

Do Regional Economies Need Regional Coordination?

Edward Glaeser
Harvard University, NBER

I. INTRODUCTION

In 1900, America was an ocean of agriculture that embedded dense urban islands. The 19th-century cities were spatially distinct. The cities often clustered around a natural advantage, such as a port, or a man-made advantage, such as a railroad stop. The cities spread throughout the continent because there was great wealth in the natural resources of the American hinterland and the inland cities served to move the wealth of the hinterland onto boats and trains for shipment east. The cities were compact units dispersed throughout the country.

One hundred years later, the economic forces that led to this urban landscape have changed. While America once thrived on its natural resources, its human capital is now central. There is little gain in having a city close to the cornfields. While transporting goods was once extremely expensive, now it is cheap and requires far less labor (Glaeser and Kohlhase 2004). While people once walked to work, now the overwhelming majority of Americans drive cars. As a result of these massive economic shifts and changes in transportation technology, the old island cities have been increasingly replaced by urban regions.

In Section II of this paper, I review the replacement of distinct cities with sprawling urban regions. I start by looking at the growth of counties in the regions around New York, Chicago, and Los Angeles and show the increasingly dispersed nature of the population. Across the U.S. as a whole, the post-war period had two different eras of population dispersal. Between 1950 and 1970, people moved from the densest counties to counties that were slightly less dense. This was the era of the early suburbs. Since 1970, population growth has been centered in areas that were radically less dense. This latter period also included the remarkable decentralization of employment that now characterizes most of the nation's metropolitan areas.

This decentralization of employment and population increasingly suggests that large regions, rather than cities or metropolitan areas, may be an appropriate unit of analysis. In Section III, I turn to the 10 megaregions identified by the Regional Plan Association (2006). While these regions have certain aspects in common, such as a relatively continuous distribution of people and firms, they also have substantial differences. For example, the high-income areas of the Northeast and Northern California that have specialized in information-intensive industries appear to be far more productive than the poor areas of the south, especially the Gulf Coast. High-cost regions have generally higher density and longer commute times.

I then turn to the growth patterns of these different megaregions. Just as with metropolitan areas or cities, we find that less dense regions have grown more quickly than denser regions, reflecting the general move to car-based, space-intensive living. We also find that warmer places have grown more quickly than colder regions, just as is the case with metropolitan areas and cities. However, unlike metropolitan areas, we find no connection between initial income and population growth, especially over the last 20 years. There is no sense that people are moving to the more economically productive places.

In Section IV, I argue that we should understand these patterns of regional growth as the outcome of differences in housing supply within the U.S. In the older, denser regions of the country, new housing has been far more limited than in the speedily growing areas of Florida, Texas, and the Southwest. While some of this difference reflects higher construction costs in the high-income regions, and some of this difference reflects less land availability, a growing body of evidence suggests that land use regulations rather than lack of land are responsible for the differences in housing supply across the U.S. The places that are not growing have chosen to artificially restrict new development while the growing areas are still friendly to new building.

Finally in Section V, I turn to the policy implications of an increasingly regional America.

I argue that there are pluses and minuses of greater regional coordination. Coordination makes it possible to internalize inter-jurisdictional externalities that are increasingly important in a regionalized world. However, like most economists, I remain enthusiastic about the diversity and competition that comes with local control. There are some areas, like economic development policy, where the benefits of competition seem to outweigh the benefits of coordination. In this area, regional control seems likely to be a mistake. In other areas, like transportation, the externalities seem massive and regional coordination seems extremely important.

Traditionally, much of housing policy has been locally controlled, although there are substantial differences from region to region. Indeed, the places that lodge land use controls in the hands of county governments appear to be friendlier to growth than the places where segregated suburbs empower homeowners to block new building. I suggest that a mixed regional-local system might have value. In this system, localities would maintain control over land use decisions, but regions would provide incentives to induce localities to make the right choices. Section VI concludes.

II. THE RISE OF REGIONAL ECONOMIES

The rise in regional economies does not mean that regions were unimportant historically. In 1900, cities still drew strength from their region, and high transport costs ensured that much trade was regional. Historically, Chicago was the hub of the Midwest and Boston was the urban core of New England. These places had a particular advantage by being physically proximate to the natural resources of their areas, and a significant amount of their trade was regional. They also served as ports of entry and exit for goods that were coming into and out of their particular regions, just as New York City served as the port for all of the United States.

The rise of regional economies does not mean that those transport linkages have become more important. Indeed, transport costs for goods have become less important and increasingly, important cities trade with the world rather than with their nearest neighbors. It is surely a mistake to see America's megaregions as independent economic entities that deal primarily with themselves, and the rise of these megaregions does not imply any reduction of connection with the rest of the country and the world.

The rise of megaregions is best seen as reflecting the changing patterns of location for people and firms. These regional economies differ from conventional cities because they are characterized by a continuous expanse of moderate density employment and housing rather than a spatial center for employment surrounded by high-density dwellings. Employment is no longer centered in Chicago's loop, but rather spread somewhat lumpily throughout the entire Midwest.

To see this, we start with the population patterns surrounding America's three largest cities—New York, Chicago, and Los Angeles—since 1950. Figures 1, 2, and 3 consider the distribution of the total population that lives in counties within 100 miles of those cities. In Figure 1, I show the share of the population that lives in New York County (Manhattan), the other four boroughs, counties that are outside those boroughs but still

FIGURE 1

Decline of the share of the population in New York County

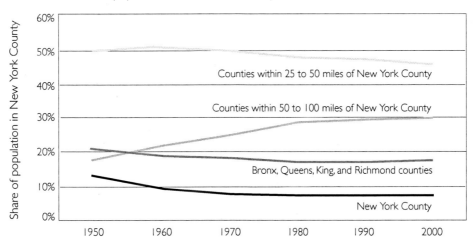

FIGURE 2
Decline of the share of the population in Cook County

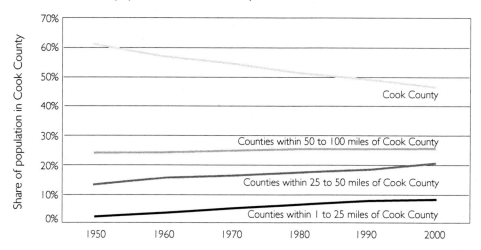

FIGURE 3
Decline of the share of the population in Los Angeles County

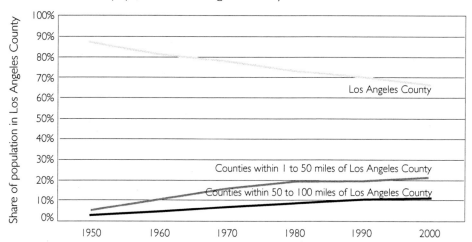

within 50 miles of Manhattan, and counties that lie between 50 and 100 miles of Manhattan. Manhattan itself declined as a share of the region between 1950 and 1970, but has held reasonably steady since then. The other four boroughs declined through 1990, but have actually risen since 1990. The inner ring of counties that are less than 50 miles of New York increased their share of population in the 1950s during the first wave of suburbanization, but have actually lost population share since then. The counties that have grown most are those on the outer edge of the city between 50 and 100 miles of Manhattan.

Manhattan is the oldest of these three cities and started in 1950 with the most decentralization, both within the city and across the inner suburbs. Moreover, Manhattan is a far smaller county than Cook or Los Angeles county so its share of population is naturally smaller. In Figure 2, I show the population distribution of the Chicagoland, again defined as counties within 100 miles of Cook County. Between 1950 and 2000, the share of the population in Cook County declined from over 60 percent of the region to less than 50 percent of the region. However, the growth in population occurred in counties between 1 and 50 miles of Cook County, not in the counties that were the furthest away. In a sense, Chicago is slightly behind New York's regional development and we should probably expect that, over time, those further counties will take a larger share of the region's population.

Figure 3 shows Los Angeles. Los Angeles is of course the most populous county in the nation and in 1950 it had almost 90 percent of the

population within 100 miles of Los Angeles. This share has declined by approximately 15 percent since then and the growth has mainly been in counties within 1 and 50 miles of Los Angeles County. As in the case of the Chicago counties, growth has been much more limited in the distant counties.

Figures 4, 5, and 6 show a different cut on the same data. They plot growth in county population between 1950 and 2000 on distance from the central county. Figure 4, the New York graph, shows a fairly clear positive relationship where the growth has been fastest in the most distant counties. Figure 5, the Chicago graph, shows that the central county itself had little growth, but once you look at Cook County, growth was actually fastest in the innermost areas. Figure 6, the Los Angeles graph, looks something like Figure 5, where Los Angeles had the least growth, but outside of Los Angeles, growth is faster in the inner counties. Again, these figures show the difference between a more mature region, where growth has been focused in the outer counties, and slightly newer areas, that are still filling up the inner circle.

Another way to see the rise of regionalism is to look at the distribution of population across counties of different density levels. In Table 1, we split U.S. counties up on the basis of density in 1950. The five rows reflect the population in counties with less than 33 people per square mile (which was the median density level of U.S. counties in 1950), counties with between 33 and 59 people per square mile (which included one quarter of U.S. counties in 1950), counties with density levels between

FIGURE 4
1950–2000 County population growth on distance from New York County

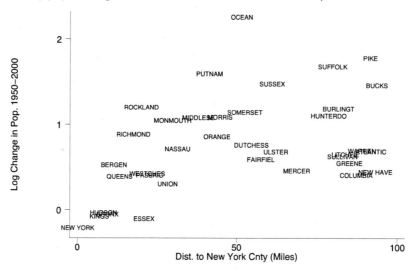

FIGURE 5
1950–2000 County population growth on distance from Cook County

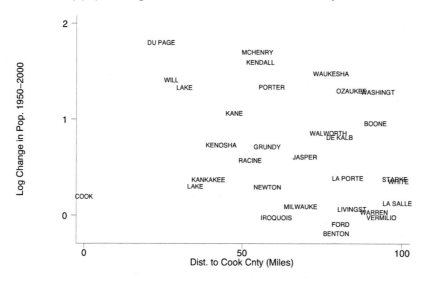

FIGURE 6
1950–2000 County population growth on distance from Los Angeles County

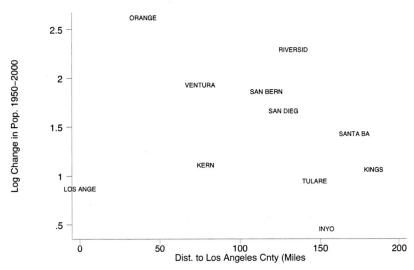

TABLE 1
Distribution of population across counties of different density levels

	Share of the population in the **least dense** counties (bottom 50th percentile) (less than 33 people per sq. mile)	Share of the population in the **low mid-density** counties (51 to 75th percentiles) (33 to 59 people per sq. mile)	Share of the population in the **high mid-density** counties (76 to 90th percentiles) (60 to 144 people per sq. mile)	Share of the population in the **dense** counties (91 to 99th percentiles) (145 to 3,670 people per sq. mile)	Share of the population in the **densest** counties (top 1 percentile) (more than 3,670 people per sq. mile)
1950	14%	13%	16%	42%	15%
1960	13%	12%	16%	46%	13%
1970	12%	11%	16%	48%	12%
1980	14%	12%	18%	46%	10%
1990	14%	13%	18%	46%	9%
2000	15%	14%	19%	44%	8%

60 and 144 people per square mile, which included 15 percent of counties in 1950, counties between 145 and 3,670 people per square mile, which included 9 percent of counties in 1950, and counties with more than 3,670 people per mile, which included the densest 1 percent of counties in 1950.

Between 1950 and 2000, the share of the national population living in the very densest counties has almost fallen in half, from 15 percent to 8 percent of the population. This decline is the essence of the decline of the old big cities. After all, of the 10 largest cities in the U.S. in 1950, eight have less population today than they did then. Those cities were the densest places 50 years ago and today they have declined substantially

While the decline of these areas has been more or less continuous, the growth of other types of counties can be clearly divided into two distinct periods. Between 1950 and 1970, those counties that had between 145 and 3,670 people per square mile saw the biggest increase in their share of the U.S. population. These were places like the inner suburbs of New York that boomed during the early post-war period. During this era, the share of population in these places grew by six percentage points, which is a lot of people, but the share of population in the least dense areas was declining. The continuing flight from agriculture meant that old farm counties were being depopulated. In total, the share of the population living in counties with less than 60 people per square mile declined from 27.2 percent in 1950 to 23.8 percent in 1970.

Since 1970, the second densest category of counties has lost ground while the less dense counties have gained. Since that year, the share of the U.S. population living in counties with less than 145 people per square mile has increased from 40 percent to 48 percent. America is becoming an exurban nation. People that once clustered in those highly dense counties are now increasingly spread out in less dense areas.

The spread of people into suburbs was followed by a spread of employers. Following Glaeser and Kahn (2004, 2001), we use zip code data on employment collected for 1994 and 2001 and characterize the degree of decentralization of employment. This data captures the number of workers—in different industries—in all of the nation's zip codes. We connect this data with information on the location of a metropolitan area's employment center, which comes from a 1982 census document that essentially established this center by polling local leaders. With this, it is possible to characterize the degree of decentralization in each metropolitan area and the changes between 1994 and 2001. It is not possible to go back prior to 1994.

Perhaps, the simplest measure of the degree of decentralization is the share of population at various distances from the central business district. Figure 7 shows the average share within 3 miles, between 3 and 5 miles, between 5 and 10 miles, between 10 and 15, between 15 and 20, and between 20 and 25 miles across all metropolitan areas for data from 2001. We ignore zip codes that are farther than 25 miles from the city center. On average, 24.2 percent

FIGURE 7
Population living and employed persons working at various distances from a central business district (CBD)

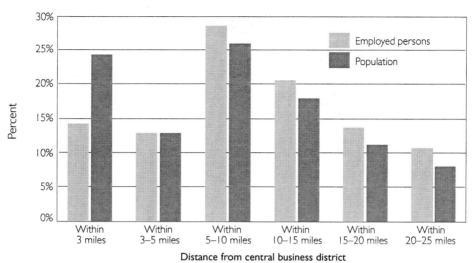

FIGURE 8
Employment centralization across MSAs

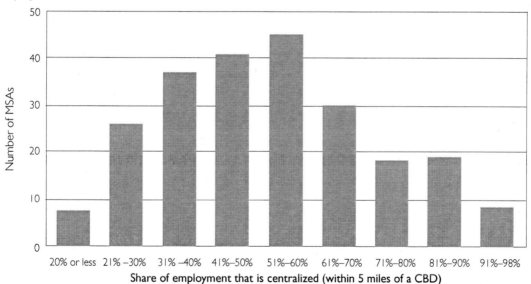

of employment is within 3 miles, 12.7 percent lies between 3 and 5 miles, 25.7 percent between 5 and 10 miles, 18.4 percent between 10 and 15 miles, 11.0 percent between 15 and 20 miles, and 8.0 percent between 20 and 25 miles. This distribution of employment confirms the remarkable level of employment decentralization that currently exists in the American city.

Of course, the average level of decentralization masks the considerable heterogeneity that exists within metropolitan areas. We use the share of employment within 5 miles of the city center as our proxy for employment centralization. Figure 8 shows the cumulative distribution of this measure across all metropolitan areas. There are some metropolitan areas with more than 50 percent of their employment that is that centralized and others with less than 20 percent of their employment within 5 miles of the city center. The mean value of this variable is 52 percent.

Glaeser and Kahn (2001) find few significant correlates of decentralization at the metropolitan area level. At the industry level, we find that high human capital industries, such as finance and insurance, have tended to remain centralized, while manufacturing is particularly decentralized. Dense areas keep their attraction for those areas that still value that ability to communicate quickly. Manufacturing has a great need for land and, with the rise of the truck, found it easy to decentralize.

CAUSES OF DECENTRALIZATION
Glaeser and Kahn (2004) argue that the primary cause of this suburbanization is

the internal combustion engine. Among individuals, cars have proven tremendously attractive because of the vast increase in speed associated with driving to work. The average car commute is 24 minutes and the average commute by public transit is 48 minutes—an enormous attraction to Americans. The truck was also important as it enabled manufacturing firms to leave the rail and port infrastructure that had tied them to cities.

The decentralized regions are essentially a redefinition of the American urban landscape around the automobile. As opposed to almost all previous transportation technologies, the car operates essentially point-to-point, rather than on a hub-and-spoke model, where the train or bus stop is essentially a hub and walking is needed to get to the final destination. The point-to-point car technology is the source of its great speed but also the reason why cars have been associated with totally different density levels. Cars both facilitate living at far lower densities and they are also huge consumers of space, which makes lower density levels necessary. The decentralization of population and employment has essentially been an attempt to accommodate the car.

Glaeser and Kahn (2004) provide a number of pieces of evidence on the role that cars played in facilitating decentralization. It is unsurprisingly true that there is a tight link between decentralization and automobiles across American cities. This connection also holds true outside of the United States. Across countries, anti-car policies, such as high gas taxes, are associated with both less car usage and denser development. To address the possibility of reverse causality (fewer cars lead to higher gas taxes), Glaeser and Kahn (2004) instrument for gas taxes using French legal origin. French legal origin countries have many regulations of most forms and also have higher gas taxes. They also have denser cities.

There are certainly other theories of decentralization that also have some merit. For example, some economists have linked decentralization to a desire to flee from a perceived blight of racial minorities. While there is surely some truth to this view, decentralization is fairly ubiquitous across American metropolitan areas and even areas with almost no minorities have large amounts of decentralization. A second hypothesis given by Margo (1992) is that suburbanization reflects rising incomes and a demand for land. Again, while there is surely some truth to this hypothesis, decentralization has occurred in rich and poor places alike and our estimates of the income elasticity of demand for land (Glaeser, Kahn and Rappaport, 2007) suggest that income effects alone would be far too small to account for much of the decentralization.

III. COMPARISONS BETWEEN MEGAREGIONS

The increasingly decentralized nature of the U.S. economy seems to call for different modes of analysis other than our traditional focus on cities and metropolitan areas. The increasingly dispersed distribution of population, at least within certain, broad regions, has led the Regional Plan Association (2006) to megaregions, which are vast areas that generally encompass many metropolitan areas. The use of megaregions reflects the fact that proxi-

mate metropolitan areas are increasingly linked because people and employment have grown between the old employment centers. In this section, I will present an empirical look at the Regional Plan Association's 10 megaregions and the differences between them. I will turn in Section V to a discussion of the potential benefits and costs of handling more government policy at the megaregion level.

The megaregions differ dramatically in income, housing costs, density, and travel patterns, both because of the current economies of these areas and because of historical factors that continue to exert a major influence. For example, higher-income areas are primarily those with a more educated workforce and longer travel times are associated with areas that historically had higher density levels. Table 2 shows the basic distribution of income, housing costs, share of the adult population with college degrees, density, and travel times across the 10 megaregions.

INCOME AND ECONOMIC DIFFERENCES ACROSS REGIONS

We have formed the average income numbers by using the Census 1 Percent Public Use Micro-Sample (IPUMS) to form the average income across all of the counties in each megaregion in 2000. Table 2 gives the range of these average income numbers across the 10 megaregions that are ranked in order of their income. The two richest regions are Northern California and the Northeast, both of which have average household incomes that are slightly above $70,000. The standard economic view is that wages reflect the marginal product of labor, so this fact implies that these

TABLE 2
Characteristics of Megaregions in 2000

	Income	Share college	Housing values	Commute time	Density level
Northeast	$70,158	30%	$133,275	29.0	801
Northern California	$70,122	30%	$176,431	26.5	265
Southern California	$61,777	24%	$133,824	27.0	352
Cascadia	$60,076	28%	$134,489	24.4	158
Midwest	$59,230	24%	$100,781	23.2	264
Texas Triangle	$58,881	25%	$ 73,967	25.7	193
Piedmont	$56,955	25%	$ 93,783	25.0	253
Arizona Sun Corridor	$56,845	25%	$100,130	24.7	93
Southern Florida	$55,563	22%	$ 93,366	25.2	397
Gulf Coast	$45,506	18%	$ 65,725	23.3	146

Note: All data from the 2000 U.S. Census

two regions are by far the most economically productive areas in the country. Alternative measures of productivity, such as country business patterns, confirm the enormous output of these two regions.

After these two megaregions there is a steep dropoff of over $8,000 and then there is a cluster of seven megaregions where average earnings run from $55,563 (South Florida) to $61,777 (Southern California). The other five regions in this group are, in ascending order of income, the Arizona Sun Corridor, Piedmont, the Texas Triangle, the Midwest, and Cascadia. All of these areas are substantially better paid than those people who live in none of the megaregions, who earn an average of $50,737, and the Gulf Coast, which earns an average of $45,506. Essentially, the megaregions consist of two high-flying areas on the eastern and western extremes of the country, one particu-

larly poor region and a large number of areas in the middle.

Across cities and metropolitan areas, two factors are known to be reliable predictors of high wages and productivity: city size (or density) and years of schooling. Ciccone and Hall (1996) document the remarkably strong connection between density and output. Glaeser and Mare (2001) show that people in metropolitan areas that surround big (and generally dense) cities earn substantially more, holding everything else constant, than people elsewhere.

Figure 9 shows the correlation across metropolitan areas between the logarithm of the gross metropolitan product per employee and the logarithm of density. The correlation is quite striking. Figure 10 shows that this correlation is also true across megaregions when

FIGURE 9
Gross metropolitan prod. per employee and population density, 2000

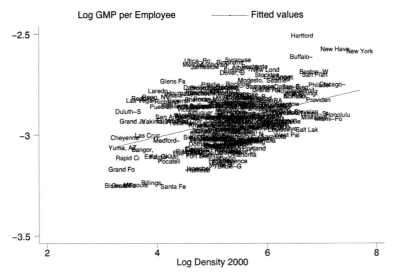

FIGURE 10
Mean income and population density, 2000

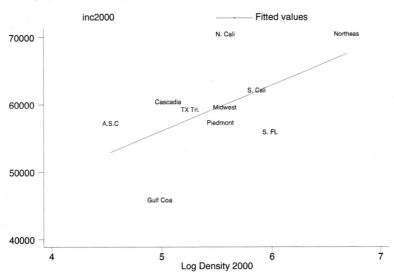

correlating mean income and the logarithm of density. The relationship is still positive although somewhat weaker. Certainly, the productive Northeast is by far the densest region and the poor Gulf Coast is among the least dense areas.

All wage differences across space require at least two explanations. First, we need to understand why firms are willing to pay higher wages in some areas than others, which means we must understand why productivity is higher in some places than others. Second, we must understand why workers don't flock to high wage areas. We will explore the second question in the next subsection, but we comment on the first question here. Dense areas may be more protective for many reasons. The omitted variable hypothesis is that locational attributes, like access to a good port, both make a place more productive and thereby attract density, which suggests that the density-productivity relationship is not the result of density causing productivity, but the result of both variables being caused by a third factor. A second explanation of the density effect is that the proximity between firms reduces the transport costs for shipping goods. A third explanation is that the close productivity of workers leads to intellectual spillovers that enhance productivity.

The third hypothesis is also an explanation for the well-known relationship between human capital and income. If areas thrive in the information age because of their ability to produce new ideas, and if agglomerations of educated people are the building material for innovation, then we should expect to see a tight link between education and regional wealth. This basic view follows the long-standing hypothesis of Alfred Marshall and Jane Jacobs who both emphasized the role of intellectual spillovers increasing productivity in dense areas.

Following Rauch (1993), there has been extensive literature showing that people who live in high human capital areas earn more. Figure 11 shows the relationship across metropolitan areas between the residual from a log wage regression, i.e., the log wage holding education, age, and gender constant, and the share of the metropolitan area that has a college degree. More educated places have higher wages. This effect has been getting stronger over time (Glaeser and Saiz 2004) and today the average wages in an area is an extremely strong predictor of the wealth of the area. This effect holds if you use long-standing variables that drive the current education level, like the density of colleges before 1940 (as in Moretti 2004).

Across megaregions, it is also true that high skills line up with high income levels, as Figure 12 shows. Just as in the case of income, there are essentially three groups of metropolitan areas. The Northeast and Northern California have the highest education levels. 30.1 and 29.8 percent of their adult populations, respectively, have college degrees. Cascadia is almost as well educated, with 28.5 percent of its population having college degrees. The Gulf Coast has the lowest levels of education by far. Only 18.2 percent of this region's population has a college degree. The other six areas again occupy a middle ground with shares of college graduates going from

FIGURE 11
Wage residual and percentage population with bachelor's degree, 2000

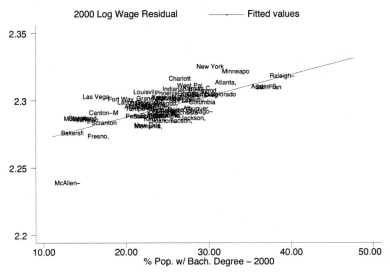

FIGURE 12
Mean income and percentage with bachelor's degree, 2000

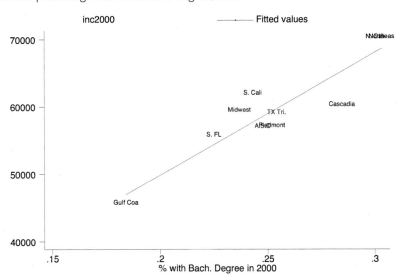

FIGURE 13
Mean income and percentage with bachelor's degree, 1950

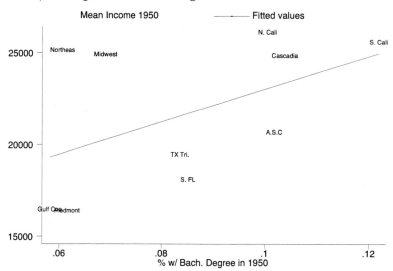

22.1 percent (Southern Florida) to 25.4 percent (the Texas Triangle).

While the basic relationship between skills and productivity certainly holds at the megaregion level, it is also clear that this relationship is far from perfect. Cascadia seems too poor relative to its education level and Southern California seems too rich. Further research is needed to understand these outliers.

Still, the relationship between skills and earnings is much stronger than it was in the past. Figure 13 shows the relationship between income and share of the population with college degrees in 1950. For 1950 income, we used the population weighted average of county incomes rather than using individual level data, but the differences between Figures 12 and 13 are not the result of this slightly different methodology. In 1950, education was mainly a function of being a newer region, with

younger people on the West Coast. Some of these places were rich, but others were not.

In 1950, the regions also had a very different distribution of income. There were essentially five richer regions: the Northeast, the Midwest, Northern and Southern California, and Cascadia. These areas all had average incomes around $25,000 (in year 2000 dollars). By contrast, the other five megaregions were all relatively poor, with the Gulf Coast and Piedmont regions at the bottom of the pack. The divergence of those two regions over the past 50 years is one of the remarkable stories of post-war American economic geography.

Another way to understand the different regional economies is to look at the primary industries of their largest metropolitan areas. The Northeast region's largest metropolitan area is, of course, New York. Ranked by payroll, finance dominates the region with

$86 billion of payroll in the Standard Industrial Classification (SIC) two-digit industry "finance and insurance." The largest subpart of this group is SIC four-digit industry "securities and commodity contracts, intermediation and brokerage." The SIC two-digit industry "Professional, Scientific, and Technical Services," has $46 billion of payroll. Health care and wholesale trade are the other two largest industries by payroll. Health care and social assistance is the region's largest employer followed by finance and insurance, and wholesale trade has the largest total receipts. The great export industries of the Northeast region are the human capital-intensive sectors of finance and business services and the more traditional residual sector of wholesale trade.

The San Francisco metropolitan area, which is the dominant metropolitan area in Northern California, has a remarkably similar industrial mix. Its dominant export industries are again finance and insurance and professional services. Northern California is also the home of Silicon Valley and its remarkable agglomeration of technology producers. Again, this thriving wealthy region is productive because it specializes in idea-intensive industries that employ highly skilled people. Seattle, the largest combined statistical area in Cascadia, is likewise an information producer. Its dominant industries are "information," which includes publishing, business services, and, more importantly, software.

Another set of five regions—the Midwest, the Texas Triangle, Southern California, Piedmont, and the Arizona Sun Corridor—all have manufacturing as their largest industrial group by payroll. These areas also have significant business services, but unlike the first three areas, they are still first and foremost in the business of producing goods, rather than ideas. The type of goods obviously differs from region to region. The Texas Triangle has the energy sector and Southern California has a vast trade associated with the port of Los Angeles, but they are basically goods producers.

The final two megaregions, Southern Florida and the Gulf Coast, have health care and social assistance as their dominant industry. In the Gulf Coast there is also some manufacturing and in Southern Florida there is some degree of business services. Southern Florida's economy is certainly also tied to tourism and retirement. The Gulf Coast is the most economically troubled megaregion.

THE SPATIAL STRUCTURE OF EMPLOYMENT ACROSS MEGAREGIONS
Not only do the megaregions differ in the amount of their earnings, but they also differ in where they earn. Some areas are relatively centralized and others are not. For each megaregion, I calculated the average share of employment within 5 miles of a central business district. There are generally several such districts within a megaregion. Figure 14 plots the shares of employment that are more than 5 miles from a central business district for all 10 megaregions and the rest of the U.S. in 1994 and 2001.

The figure shows that all megaregions are more decentralized in 2001 than in 1994, which can be seen in the fact that all of the data points are below the 45 degree line

FIGURE 14
Employment within 5 miles of CBD, 1994 and 2001

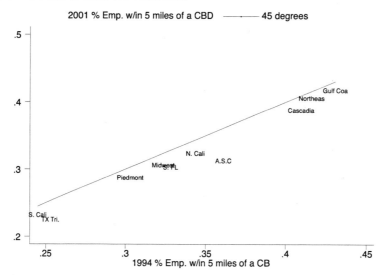

FIGURE 15
Mean housing value and mean income, 2000

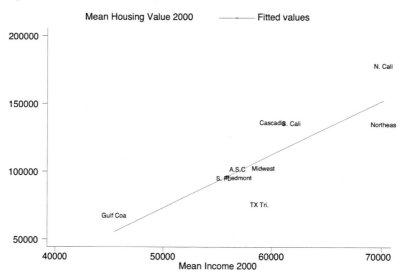

shown in the figure. In most cases, the changes are modest, but this is, after all, only a seven-year time period. If anything, the Arizona Sun Corridor has had the biggest increase in decentralization, but even its change is only about 5 percent.

Across the areas, there are three clusters. First, the Gulf Coast, the Northeast, and Cascadia are the most centralized. In these cases, around 40 percent of employment is within 5 miles of a central business district. This is the pattern that holds outside of the megaregions as well. The second cluster contains five megaregions: the Arizona Sun Corridor, the Midwest, Northern California, Southern Florida, and Piedmont. In these places about 30 percent of employment is within 5 miles of the central business district. As such, these places are essentially hybrids of older centralized areas and more modern job sprawl. Finally, the Texas Triangle and Southern California have about 23 percent of their jobs within 5 miles of a central business district. These places have very little centralization.

A finer look at the data shows that the 5-mile cutoff misses some interesting variation across the megaregions. If we look within the 5-mile area, the Northeast, Cascadia, Midwest, and Northern California have considerably more employment within 3 miles of central business districts than between 3 and 5 miles of the district. In the other regions, the distribution of employment is relatively flat within these 5-mile areas. As such, some of these areas really continue to have true employment centers while others do not.

HOUSING COSTS AND OTHER DISAMENITIES

The high-wage megaregions are also high cost areas. After all, if they weren't then everybody would want to live in places where they could earn more money and, as a result, nobody would want to live in the Gulf Coast. Economics has no more fundamental principle than that high wages should be offset by other costs, as seen in the distribution of housing prices across megaregions.

Figure 15 shows the relationship between housing prices and incomes across megaregions. Our housing value numbers are based again on the 2000 Census IPUMS and we are forced to use self-reported housing values. The basic relationship between prices and income is quite tight, as it is across metropolitan areas. Figure 16 shows the fact that metropolitan area income is a strong determinant of housing values across metropolitan areas. On average a $10,000 increase in income is associated with a $37,000 increase in housing values in 2000.

The major outliers in the cross-regional relationship between income and housing prices are Northern California, which is more expensive than its income seems to merit, and the Texas Triangle, which is cheaper than its income would seem to merit. The most natural explanation for these differences is climate and other amenities. As Figure 17 shows, there is a robust positive relationship between warm Januaries and housing prices across metropolitan areas. Northern California may offer the same income as the Northeast, but it also offers a much more moderate climate.

FIGURE 16
Housing value and income across MSAs, 2000

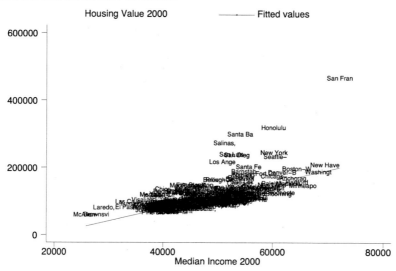

FIGURE 17
Housing value and mean January temperature, 2000

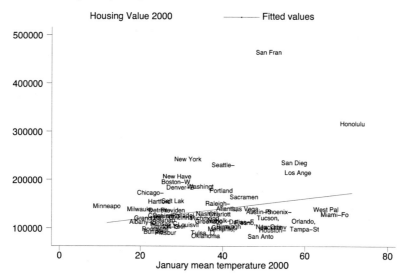

FIGURE 18
Commute time and mean income, 2000

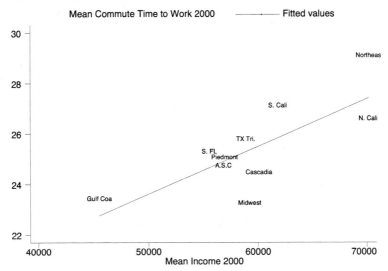

Unsurprisingly, people are willing to pay for this comfort. Conversely, the Texas Triangle has a particularly tough climate that is both extremely hot and quite humid. Perhaps these variables explain the low price of Texas homes.

As a result of the twin forces of climate and income, megaregions can be divided into four areas on the basis of housing prices. Northern California is in a world of its own with median housing values of more than $175,000. These values have risen dramatically since 2000. The Northeast, Cascadia, and Southern California form a middle group with prices hovering around $130,000. The Midwest, the Arizona Sun Corridor, Southern Florida, and Piedmont are a fourth group with prices between $90,000 and $100,000. Finally, the Texas Triangle and the Gulf Coast are much cheaper than the other regions.

The differences in housing prices reflect both housing demand and housing supply. The four most expensive places are also places where new construction has been limited. In a series of papers, Joseph Gyourko and I have argued that these limits on supply are primarily the result of land use regulations. I will return to these issues of housing supply at the end of the section.

Housing costs only capture one of the prices that must be paid for living in expensive areas. The more productive, dense regions also have longer travel times. Figure 18 shows the relationship between average commute time and income across metropolitan areas (megaregions). People in the high-income regions not only pay more for their houses, but they also must spend more time commuting. This is a classic prediction of the Alonso-Muth-Mills model where people can obtain cheap housing on the urban fringe. As a city gets more

FIGURE 19
Commute time and population density, 2000

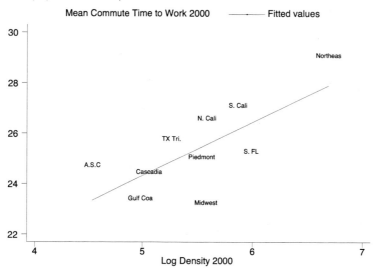

Mean Commute Time to Work 2000 ——— Fitted values

productive, prices go up, but people also live further and further away from the city center and have longer and longer commute times.

In the figure, the Northeasterners spend more time commuting than their income would suggest, while the Northern Californians have a somewhat shorter commute. This presumably is one of the reasons, along with climate, that housing prices are higher in Northern California. Southern California has somewhat higher commute times than its income might suggest and this might be seen as another way in which Los Angelenos are paying for their famous weight. Commute times in the Texas Triangle, Southern Florida, Piedmont, Arizona, and Cascadia are all fairly similar and close to the regression line. Cascadia does appear to have somewhat shorter commutes than its income might predict, but then again, its housing prices are also higher than its income might predict.

Finally, the Midwest and the Gulf Coast have much shorter commute times. In the case of the Gulf Coast, the low commute time is predicted by the low incomes. One advantage of living in a place with a weak economy is less traffic on the roads. The Midwest has the lowest commute times of all of the megaregions. This is clearly one form of compensation for cold Midwestern winters.

While it is useful to think about commute times as part of the price that people have to pay to live in economically robust regions, density, rather than income, is a more direct cause of long commutes. Figure 19 shows that relationship between density and average travel times to work across megaregions as it is across cities. Unsurprisingly, more people per square mile means slower commutes. Figure 20 shows the relationship between density and commute times across cities with more than 100,000 people and more than

153 people per square mile. Some of this relationship occurs because people in dense areas are more likely to use public transportation, and public transit commutes are generally much slower than car commutes. Figure 21 shows the relationship between commute times and share of the population using public transit using the same sample as Figure 20. In a regression where commute times are regressed on both public transit use and density, both variables significantly increase commute times.

THE GROWTH OF MEGAREGIONS

The megaregions differ not only in their current characteristics but also in their patterns of growth. Some of these regions are growing strongly; others are losing ground relative to the U.S. as a whole. Figures 22 and 23 show the share of the U.S. population living in the 10 megaregions over the past 50 years. Figure 22 shows the share living in no megaregion, the Northeast, and the Midwest. These three groups have many more people than the other megaregions, so it is easiest to consider them separately. All three lines are on a downward trajectory.

From 1950 to 1970, these two regions roughly held their ground. Since 1970, the regions have lost population relative to the rest of the nation. About 10 percent less of the U.S. population lives in these two regions compared to 30 years ago. It is no surprise that these older, rustbelt areas have lost population. After all, warm temperature is among the most reliable predictors of urban growth over the past century (Glaeser and Shapiro 2003).

I have trouble concluding too much from the decline in the share of the population not living in any megaregion. After all, the megaregions are defined on the basis of their current importance to the U.S. economy. If these areas had been declining, they would have been less likely to have been selected with that criterion.

Figure 23 shows the rise of almost all of the other areas. Southern California and Southern Florida have had the greatest increase in their share of the U.S. population. Both of these regions have seen a roughly 5 percent increase in their share of the U.S. population. The increase is particularly spectacular in Southern Florida, which started the post-war period with 1.5 percent of the U.S. population. Southern Florida's rise owes much to immigration and the growing retired population, but there has also been an impressive increase in native population of all age groups. Rising incomes have led people to increasingly value the consumer amenities of this semi-tropical area.

The Texas Triangle and the Arizona Sun Corridor have also had impressive growth. As in the case of southern Florida, Arizona's growth is particularly impressive because it began with such a small share of the U.S. economy. The growth of Cascadia and Piedmont is also impressive, but it is somewhat more modest than the other growing regions. Finally, as one might expect from its economic weakness, the Gulf Coast has had a relatively constant share of America's growth over the past 50 years.

What explains the growth of the different regions? Three main factors have historically been important for the growth of metro-

FIGURE 20
Commute time and population density, 2000

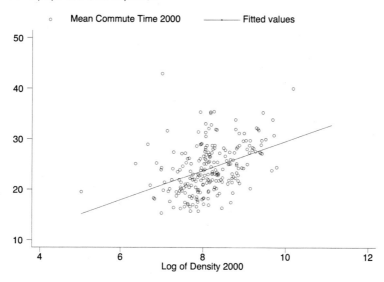

FIGURE 21
Commute time and percentage using public transportation, 2000

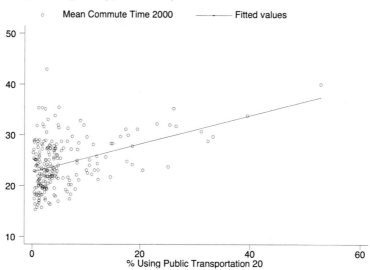

FIGURE 22
Share of population in the Northeast and Midwest Megaregions, 1950–2000

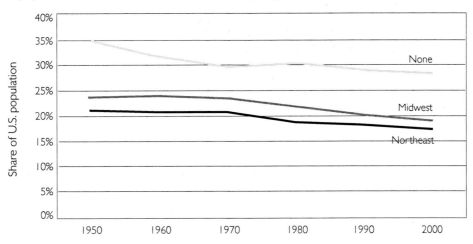

FIGURE 23
Share of population in the Arizona Sun Corridor, Cascadia, Gulf Coast, Northern California, Piedmont, Southern California, Southern Florida, and Texas Triangle Megaregions, 1950–2000

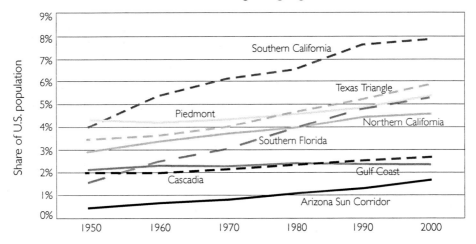

politan areas: sun, skills, and sprawl. The correlation between median January temperature and population growth between 1980 and 2000 for the 100 most populous metropolitan areas in 1980 is shown in Figure 24. The relationship between skills, as measured by the share of the population with college degrees, and population growth is shown using the same sample over the same time period in Figure 25. This skills connection is closely related to the connection between economic success and subsequent growth. Figure 26 shows that this correlation is much stronger in the older, colder regions of the country. Finally, Figure 27 shows the relationship between density and growth over the same time period.

Across megaregions, both density and warmth predict growth. Figure 28 shows the correlation between density in 1950 and population growth across megaregions since then. It is

clear that the less dense places have had faster growth, which may not be surprising since they are starting from a lower base. Figure 29 shows the positive connection between average January temperature and growth in 1950 and since then. Less density and more sun seem to predict growth at the megaregion level, just like they do at the metropolitan area level.

However, the relationship between income and megaregion growth shows that people are not moving to richer areas. Glaeser and Shapiro (2001) find that in the 1990s, initial income predicted the growth of cities and metropolitan areas, but the richer megaregions have not grown faster, perhaps because those richer places are also denser. Figure 30 shows the relatively flat correlation between income and growth during the 1950 to 1970 period. Figure 31 shows the remarkable negative correlation between income and growth between 1980

FIGURE 24
Population growth and mean January temperature, 1980–2000

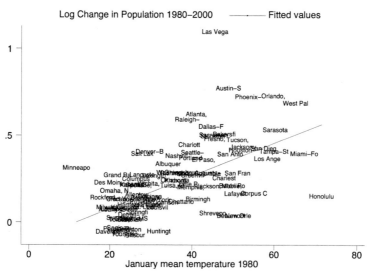

FIGURE 25
Population growth and percentage bachelor's degree, 1980–2000

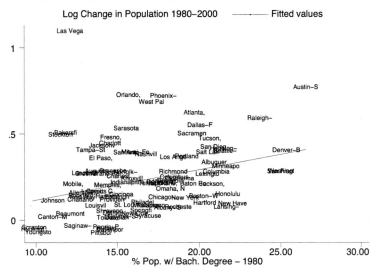

FIGURE 26
Population growth and percentage bachelor's in Northeast and Midwest, 1990–2000

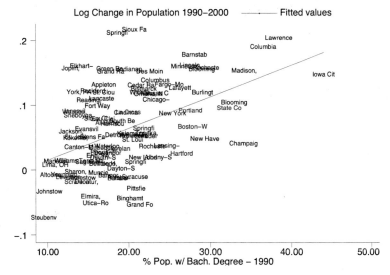

FIGURE 27
Population growth and population density, 1980–2000

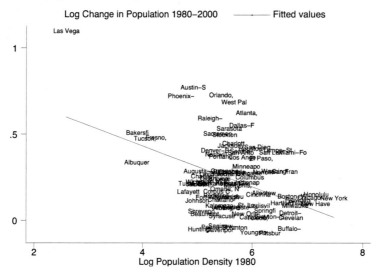

FIGURE 28
Change in population and 1950 population density, 1950–2000

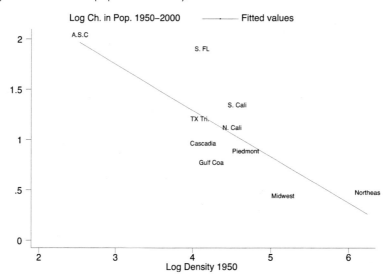

FIGURE 29
Change in population and mean January temperature, 1950–2000

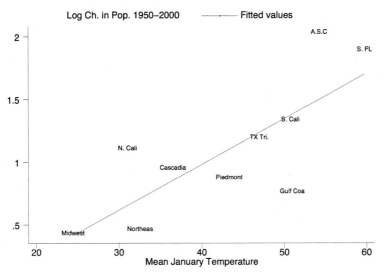

FIGURE 30
Change in population and 1950 income, 1950–1970

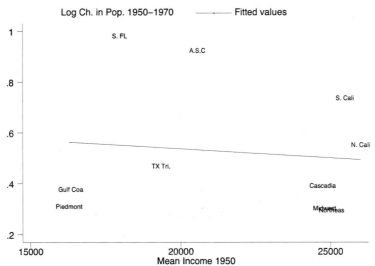

FIGURE 31
Change in population and 1980 income, 1980–2000

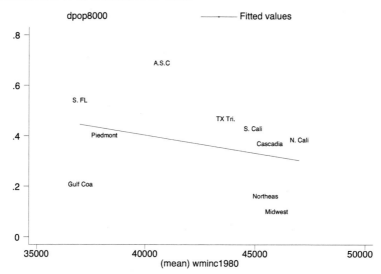

and 2000. Over the last 30 years, economic growth has been concentrated in the poorer megaregions.

In the next section, I will argue that we must understand housing supply to understand the fact that poorer megaregions are growing much more quickly than rich megaregions, and that the connection between housing supply and regional growth is one reason why it makes sense to think about a more regional or national approach to land use regulations.

IV. REGIONAL GROWTH AND HOUSING SUPPLY

The fact that high-income places have grown less than low-income places would be hard to understand without understanding the importance of housing supply. If housing could be freely supplied at a fixed price with a fixed commute everywhere, then people would

presumably be flocking to high-income areas. Perhaps, some high-income areas, like the Northeast or the Midwest, lack other amenities, like climate, but that certainly can't be said of Northern California. Yet Northern California, despite its high incomes, splendid climate, and relatively short commutes, grew less than five of the 10 regions over the past 20 years.

Over the past 20 years, the explosive growth was in the Arizona Sun Corridor, the Texas Triangle, and Southern Florida. The five faster growing metropolitan areas in the 1990s were Las Vegas; Naples, Florida; Yuma, Arizona; McAllen, Texas; and Austin, Texas. These places are generally not as economically successful as coasts and their climate is certainly worse than that of California. Their growth has more to do with housing supply than with innate demand for the characteristics of these areas.

A basic fact about population growth is that change in the number of people and change in the housing stock are almost perfectly correlated. Figure 32 shows the correlation between the change in the number of homes and the change in the number of people in the 1990s. Over very short time horizons, there can be changes in the vacancy rate and over longer time horizons there has been a secular decline in the number of people occupying each housing unit. However, when we compare metropolitan areas or cities in a given decade, growth in population and growth in the housing stock are almost the same thing.

This fact would be relatively uninteresting if all areas had roughly the same housing supply and the differences in both population and housing change were driven by changing demand to live in a particular region. For example, if some regions were attractive because of income

and amenities and others were not and if growth was determined by these factors, then we would still expect to see the correlation between income and housing growth, even if housing supply was perfectly elastic everywhere. However, this view of the world would also predict that prices should be higher in places that were growing more. After all, if growth is driven by demand, not supply, then high demand should lead to both high quantities and high supply.

Figure 33 shows the relationship between population growth between 1990 and 2000 across megaregions and housing prices in 2000. The growth does not show any sort of a straight line connecting housing prices with growth. Some of the data points in Figure 33 can be explained by demand. The Gulf Coast, and to a lesser extent the Midwest, combine low prices and low growth. These areas do

FIGURE 32
Change in housing units and population, 1990–2000

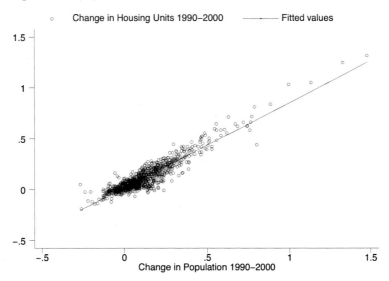

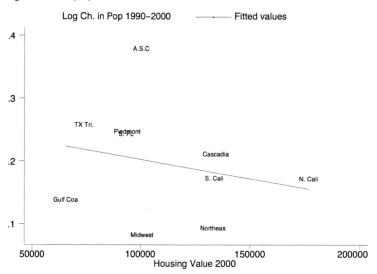

FIGURE 33
Change in housing value and population, 1990–2000

seem to have less demand. However, among the other areas, high growth seems to be correlated with low prices, not high prices. The Texas Triangle is one of the faster growing megaregions and yet its prices are almost as low as those in the Gulf Coast. Northern California and the Northeast are among the slowest growing regions and they have the higher housing prices.

These patterns can also be seen at the metropolitan area level, as shown in Figure 34. This figure shows permits issued between 2000 and 2005 and housing prices in 2000 across metropolitan areas. The graph shows three basic sectors. First, there are areas with few permits and low prices. These are low demand areas, many of which are in the Midwest and the Gulf Coast. Second, there are areas with many permits and moderate prices. Most of these areas are in the fast-growing megaregions. Third, there are areas with few permits

and high prices. Many of these are in the Northeast and Northern California.

The fact that high-price areas have little new permitting and high-permitting areas have moderate prices is certain evidence of supply differences across areas. Without differences in supply conditions, the expensive areas would be building much more than the cheap areas. Instead, we see that building is going on in areas with less demand because in those areas it is easier to build housing. The constant flow of new homes in these places helps to ensure that they stay affordable.

LACK OF LAND VS. LAND USE REGULATION

There are essentially three ingredients in building new homes: land, a permit, and physical infrastructure. In principle, supply might be greater in some areas than others because of any of these three factors. High supply places

FIGURE 34
Prices in 2005 and permits 2000–05 across MSAs

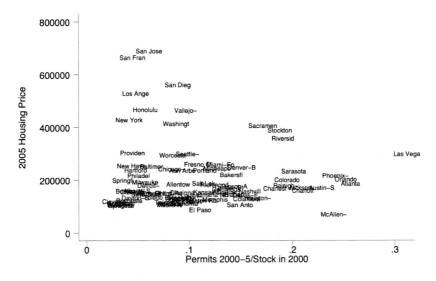

might have more land, or easier permitting processes, or better provision of housing structures. In this section, we will discuss the evidence on the relative importance of these three inputs to differences in supply conditions in metropolitan areas.

Gyourko and Saiz (2004) have looked at the differences in the cost of building structures across space using data from R. S. Means that surveys builders on their building costs. They do indeed find that physical construction costs are higher in areas with high housing prices; however, differences in physical construction costs can only explain a small amount of the heterogeneity in prices. For example, they document a $25 per square foot difference in construction costs between Atlanta and San Francisco, which is among the most extreme differences in construction costs. This would explain a $62,500 difference in prices for a 2,500 square foot house. By contrast, the

National Association of Realtors reports a median sales price in Atlanta of $167,000, which is $550,000 less than the median cost of $715,000 in San Francisco. While differences in physical construction costs are not irrelevant, they can only be a small part of the story.

The more difficult task is to determine the relative importance of lack of land and difficulty in permitting. One way to look at the importance of density is to see whether there is more construction in places with more land. Figure 35 shows the correlation across metropolitan areas between permits per acre between 2000 and 2005 and density in 2000. The figure shows a strong positive correlation. Figure 36 shows a similar positive correlation across 187 cities and towns, in the greater Boston region. Places that have less land build more, not less. This result does not reflect high demand for those areas, as the correlations are unchanged if we control for initial price in the area.

FIGURE 35
Permits 2000–05, area in 2000, and density in 2000

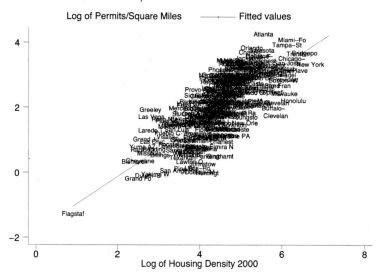

FIGURE 36
Permits 1980–2002 per acre and 1980 housing density in greater Boston

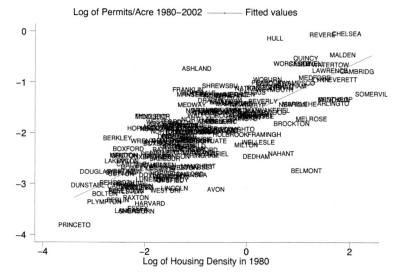

A second piece of evidence on this question is that prices are not particularly higher in areas with more density. Glaeser and Gyourko (2003) look at high prices across metropolitan areas, and find that many of the places with the highest prices have quite low-density levels. The combination of high prices, low permitting, and low density is particularly clear in Northern California, which has protected great tracts of land.

A third piece of evidence on lack of land is that lot sizes are getting bigger in at least some high-cost areas. Jacobovics (2006) documents that in the Boston area "the median lot size for new single-family houses was 0.91 acres, up from 0.76 between 1990 and 1998." This increase in lot size is far too big to be accounted for by the rise in incomes. If greater Boston was running out of land, then we would expect lot sizes to be getting smaller rather than larger. The fact seems far more compatible with the view that regulations restricting new construction are getting tougher over time.

A final piece of indirect evidence of the importance of land scarcity compares the price of land evaluated in two different ways. Conventional hedonic analysis can deliver a price of land by comparing similar houses on smaller and bigger lots. This should deliver people's willingness to pay for more area. A second approach to estimating the price of land is to look at the total sales price of a house and subtract the estimated price of building the physical structure. This second approach will combine both the value of the land and the value of having the right to build on that land.

In an unregulated market, these two methods of valuing land should yield the same result. Areas with high demand and scarce land should have equally high land values using either technique. In an unregulated world, a quarter-acre would be worth more if it sat under a new house, and homeowners would subdivide their lots and build. Glaeser and Gyourko (2003) compare these two different methods across many metropolitan areas. We find that in high-cost areas the price of an acre is typically 10 times larger if it sits under a new house than if it extends a lot. This finding is incompatible with the view that these places just lack land, and so there must also be substantial regulations preventing subdivision.

In a related exercise, Glaeser, Gyourko, and Saks (2005) examine prices and construction costs in Manhattan. If Manhattan's limited supply and high prices just reflected a lack of land, then we should expect new residential buildings to be getting taller. By contrast, they are getting shorter. If regulations didn't bind the market, then we should expect to see apartment prices roughly equal the physical cost of building up. After all, with residential high rises, you don't need more land to create new apartments; you just need to build taller buildings. We find that condominium prices are now at least double construction costs in many areas. Again, this is incompatible with the view that lack of land is the only restriction on new construction.

There is also more direct evidence on the role that growth controls, minimum lot sizes, and other forms of regulation play in limiting the amount of new development. Katz and Rosen

(1988) wrote a classic paper about the Northern California region that shows that prices are higher in areas that have used regulations to restrict growth. Across metropolitan areas, Glaeser and Gyourko (2003) find that prices are higher in more regulated markets.

Glaeser and Ward (2006) look at land use regulations within the greater Boston region. We document a remarkable panoply of regulations that impact the ability to build, including rules about subdivisions, wetlands, and septic systems. These rules vary wildly from place to place for reasons that seem to be unrelated to any obvious economic or physical factor. These rules have also increased significantly over time.

We examine the connection between land use regulations and both new construction and prices. We find that as average minimum lot size in a town increases by one acre, the number of new permits between 1980 and 2002 decreases by 0.41 log points or roughly 40 percent. When towns adopt an extra type of land use regulation, we find that new construction drops by 0.105 log points or roughly 10 percent. Each extra acre of minimum lot size appears to increase prices by about 15 percent and each extra form of regulation increases prices by about 10 percent. Land use regulations appear to both decrease new construction and increase prices.

I do not mean to suggest that land availability is unimportant. Surely, some part of the remarkable growth of the Texas Triangle and the Arizona Sun Corridor comes from an abundance of land. Yet regulations are also extremely important. The importance of these regulations means that the housing supply differences across space are not fixed, but rather could be changed under a different regulatory regime.

THE POLITICAL ECONOMY OF ZONING AND URBAN DECENTRALIZATION

The popularity of land use controls and their rise over the last 35 years is not hard to understand. For most homeowners, new construction is a bad thing. There are both real externalities and pecuniary externalities from every unit of new construction. The real externalities include congestion on the roads and the potential nuisance of actual construction. Many people may prefer to be surrounded by land rather than homes so there is a real aesthetic loss associated with new building. If new homes don't pay enough in taxes to cover the expected costs of providing them with public services, then there may also be an externality working through the public treasury.

The pecuniary externality is that each new home increases the supply of housing, which should lower the price of housing. Just as OPEC has an incentive to restrict the supply of oil, local homeowners have the incentive to restrict new housing to maximize the value of their own house. Particularly in high-cost areas, homes are often a huge part of a household's portfolio. We shouldn't be surprised that homeowners are willing to work hard to restrict the supply of substitutes for their most valuable asset.

In a conventional city, the interests of homeowners were opposed by the interests of employers, builders, and bankers, who together made up what Logan and Molotch (1988) called the "urban growth machine." Employers want to make sure that housing prices are low so that they will need to pay their workers less. Builders profit from construction and have an incentive to push for more permits. Finally, bankers earn profits from lending to new homebuyers and also have an incentive to encourage new construction. Historically, this group has acted to make construction easy and these actors are still important in the high-growth megaregions such as the Texas Triangle and relatively pro-growth cities like Chicago.

If traditional cities pitted homeowners against this growth machine, the pro-growth elements are remarkably absent from many suburbs. Bedroom suburbs are homeowners' enclaves with little employment. Their governments are dominated by their major interest group—the homeowners who make up the population. We should not be surprised, therefore, that in many places where homeowners have control over governments, that those homeowners have made it increasingly difficult to build new housing.

This logic can help us make sense of the connection between urban decentralization and the rise of land use regulation. In the older centralized cities, all of the different groups interacted and often the large employers had the upper hand. As the car pushed people out into specialized suburban enclaves, homeowners came to dominate the decision-making process and shut down new development, particularly in high-income, high-education areas.

Of course, this process did not happen everywhere. In Boston and Northern California, anti-growth groups were particularly effective in blocking new construction. In the growing megaregions, these groups have been much less effective. One set of reasons for these differences are differences in long-standing political institutions. In some areas, small ex-urban communities control development. In other areas, county governments have control over new building. For example, much of the growth in Las Vegas has occurred in unincorporated areas where only the county board holds sway. Even in some eastern areas, like Maryland, county governments are the ultimate arbiters of new development and they have generally been much more supportive of new growth than small suburban enclaves.

The reason that larger governments seem to be so much more pro-growth is probably that those governments are more likely to be influenced by important employers and builders. While local homeowners may be effective in controlling a small local zoning board made up of their neighbors, they will be much less effective in pushing a county government that also has an interest in attracting employers. For this reason, larger governmental entities, such as those that have tended to dominate in the growing megaregions, have usually been more supportive of permitting.

THE WELFARE EFFECTS OF LAND USE REGULATION

Nothing that I have said in the previous discussion suggests that land use regulation is either good or bad. There are real externalities associated with new construction and it is appropriate for communities to impose some limits on completely unfettered growth. Showing that land use controls are important does not imply that they are bad. In this subsection, I will discuss some of the theoretical and empirical arguments about the welfare consequences of land use controls.

Standard economic arguments suggest that the right level of land use controls should essentially impose a "zoning tax" that is equal to the real externalities created by new development. This "zoning tax" is essentially the gap between the current housing price and the cost of supplying the house in the absence of regulation. Glaeser, Gyourko, and Saks (2005) estimate that this "zoning tax" represents about 50 percent of the value of a house in New York City and in several other high cost areas. If this estimate is correct, then new housing is essentially being taxed at a 100 percent rate because of limits on development.

Is this tax appropriate given the externalities that new housing creates? Glaeser, Gyourko, and Saks (2005) discuss the possible negative externalities associated with construction in New York. Even including the potential losses from lost views, we find it implausible that a 100 percent tax is appropriate given the positive and negative effects of new high rise construction in Manhattan. While our

estimates are certainly rough, it seems difficult to justify the limits on new construction that appear to be in place in New York City.

Similar calculations for suburban development are much more difficult because it is hard to consider the full range of changes in a suburban community that may come from new building. There is a somewhat simpler approach to this question, however, which asks whether the current level of density maximizes the total land value in the community. With a number of assumptions, land value maximization ends up being equivalent to welfare maximization. Glaeser and Ward (2006) find that current density levels in Greater Boston are far too low to be land value maximizing. Extra housing does have a negative impact on prices, but this impact is too small to offset the extra value created by building a new house. As a result, the restrictions on housing appear to be reducing land values in the area.

If these results are right and communities are building too little, then we should ask why homeowners are unable to coordinate and maximize land values. For such coordination to work, new builders would need to bribe neighbors to let them build. In some cases, we do see impact fees that resemble such bribes, but on the whole there is much less of these side payments than we might expect. In some cases, there are institutional rules that explicitly forbid such transfers. My own view is that some combination of cultural norms and laws have made these transfers rare, and as a result, homeowners only see a downside from new construction and act to prevent it.

The impact of restrictions on new construction has generally not been to limit the amount of building in America but rather to push that building from one region to another. The regional redirection of construction from highly productive regions to less productive areas may actually be quite costly for the country. Moreover, since some externalities may actually be worse when building in the growing regions, the net effect of this redirection may be negative. For example, new construction in Manhattan leaves much less of an environmental footprint than new construction on the fringes of Las Vegas. Yet our current land use policies encourage builders to avoid the city and head for the desert.

V. ECONOMICS, PUBLIC POLICY, AND MEGAREGIONS

One view, usually espoused by planners, is that the United States is hampered by having political units—cities and counties—that are too small to make investments that would benefit the country as a whole. Another view, more commonly taken by economists and political scientists who are followers of Charles Tiebout, is that competition among governments is a great way to generate both discipline and innovation. The answer is surely between those two extremes. In the case of activities where local actions have vast externalities, it surely makes sense to move towards more regional control. In other cases, where externalities are much smaller or where competition is more vital, more local control makes sense.

In this section, I will discuss three different types of public policy—economic development, education, and transportation—before focusing on housing. I argue that these three policy areas span the spectrum of areas that should have continued local control to areas that desperately need more regionalism. I ignore many important areas of governmental action, such as police, fire, and redistribution, because they seem less likely to be subjects for debate. In some of these cases, such as police and fire, I think that the case for continued local control is extremely strong because the advantages of local knowledge and competition and the modest interjurisdictional externalities. In the case of redistribution, I think that national control makes more sense both because of externalities coming from the mobility of the poor and because I think that any obligation to take care of the poor operates at the least on the national (if not the world) level.

ECONOMIC DEVELOPMENT

One of the areas where regionalism has been discussed is in the area of economic development. Since any large-scale employer will surely hire workers and buy inputs from throughout a region, it seems to make sense to run economic development policy at the regional level, rather than the local level. After all, when Boeing comes to Chicago, it will also hire people who live in Evanston. Surely it makes sense for the entire region to chip in to attract these large employers.

While this logic may seem attractive, I think it is mistaken because I think most local economic development policy is also mistaken and that those errors will tend to grow bigger at the region level. The starting point is a question about what makes for good economic development policy. One view argues for large-scale pro-action, where governments go out and lure businesses with tax cuts and other subsidies. The other view argues that attracting business is mostly about making a place attractive to workers and getting rid of governmental barriers to innovation and firm location, such as taxes and regulation. According to this second view, Nevada's laissez-faire approach has been a far more effective local economic development policy than the cities that have combined more taxes and regulation with aggressive courting of businesses.

One way of seeing the two different poles of economic development policy is that the first activist approach is really an employer-based policy that emphasizes providing stuff for firms. The second approach is essentially a person-based policy that emphasizes making a place attractive for workers and trusting that firms will follow, as long as there aren't too many barriers to business. The second approach is not laissez-faire in general. Making a place attractive to residents requires a great deal of government effort in education, public safety, transportation, and housing. However, the second approach is more laissez-faire in the areas that are commonly called economic development policy.

The case against an activist economic development policy is that governments are inherently bad at picking firms and industries that are a good investment, both because picking winners is inherently hard and because governments frequently face incentives to do the wrong thing. Japan's Ministry for International Trade and Industry (MITI) aggressively followed the path of active economic development. While some early commentators suggested that MITI was part of Japan's economic success, Beason and Weinstein (1996) show clearly that MITI targeted support to low-growth sectors that had few returns to scale. If MITI, which had vast resources and their pick of the smartest Japanese graduates of their top universities, was unable to pick winners, how can we possibly expect American local governments, with far less manpower and resources, to do better?

Certainly the track record of local economic development policies is far from encouraging. There has been a strong tendency to focus on industries of the past rather than industries of the future. Advocates of Boston-area economic development policy were arguing as late as the 1960s that without subsidies to shoe and candy manufacturing, the region would have no chance. There are many cases where subsidies to particular firms seemed to reflect political clout more than economic value.

The difficulty of micro-managing economic development should actually be getting worse in an increasingly innovative era. The essence of an innovative economy is unpredictability. After all, if we could predict the new, new thing, then entrepreneurs wouldn't be able to thrive by coming up with new ideas. We may know that biotechnology is likely to be important, but there is an entire industry trying to figure out what areas of biotechnology will succeed and cities are unlikely to be able to win at that game.

None of this is meant to say that all economic development policies are bad. Reducing taxes and regulation seems to attract business. Holmes (1998) uses a spatial discontinuity approach to show how firms located in states that had anti-union right to work laws or lower taxes. Greenstone and Moretti (2003) find that localities that are able to attract big industrial plants, usually by reducing taxes, seem to do reasonably well. One of the reasons for the great success of the Las Vegas area is surely its low taxes and lack of regulation.

If we think that good economic development policy is mostly the reduction of taxes and regulations, then local competition seems much more attractive than collective regional monopoly. There is surely some room for cooperation. Evanston may want to contribute to the tax relief offered by Chicago, but as long as we think that attracting businesses means eliminating barriers, then there is little need for all that much coordination. Localities can get rid of those barriers on their own and competition among localities provides them with the right incentive to do just that. A larger scale of government is less likely to be nimbly able to respond to the need to change the rules to attract firms.

A second factor cutting against regionalism in economic development policy is that much of good economic development policy involves looking for low-cost ways to take care of the pressing needs of local businesses. While we always need to be wary of too much

government-business cooperation because of fears of corruption, there is surely much value in having open lines of communication so that the government can relieve cheap problems facing local firms, like a traffic snarl or an environmental hazard. These actions require communication and small local governments are more likely to find that communication easy than a large regional economic development team.

A third advantage of local economic development policies is that their more limited resources make subversion by businesses less costly. A business that figures out how to influence the national government has the possibility of receiving vast federal subsidies. Localities are limited in what they can give and this limits the scope for abuse. Overall, I think that economic development policy is an area in which regionalism is unlikely to be all that helpful. Local competition seems more likely to yield attractive results than big regional planning.

EDUCATION
Education occupies a middle ground where there are surely some significant advantages for regional policy, but there are also big gains from local control and competition. There are at least three good arguments for some form of regionalism in education. First, people who are educated in one locality will often work elsewhere in the megaregion or the country as a whole. Second, people who are attracted by a good education system in one jurisdiction will often play a role in the economy of the region outside of that jurisdiction. Third,

education spending is often the best form of redistribution available because human capital is so important to generating income. As I argued above, redistribution is surely not a local responsibility.

A final fourth issue with local provision of education is that it provides powerful distortions to the location decisions of parents. Even if all communities are otherwise identical, the localization of schooling provides strong incentives for parents who care about schooling and are able to pay for it to locate together. This creates unnecessary segregation on the basis of skills and income. The problem is worse if communities are not identical and some places are bedroom suburbs while others are mixed-use cities. In that case, local schooling creates an unfortunate incentive for parents to flee the older cities with problematic school systems.

On the other hand, there is some evidence suggesting the competition among school districts improves school quality (Hoxby 2000). School finance equalization schemes have often had the impact of reducing school quality by eliminating the incentives or ability of some jurisdictions to spend more (Hoxby 2001). The track records of large school districts, which are often extremely bureaucratic and static, are far from encouraging.

As such, education is an area that has both strong advantages from regionalism and disadvantages from cutting back on local competition. There are no easy answers to this problem, especially since the current system is so politically entrenched. Teachers' unions are both enormously powerful and generally hostile to changes that would introduce competition and accountability into the system.

In such a setting, the economists' dream would look something like a megaregion-wide voucher program with lots of competition among public and private providers and no fixed connection between location and school. In such a system, the voucher could be used to handle redistribution and to ensure robust public investment in education, but there would still be the virtues of competition and diversity. As such, the system would be a hybrid that combined the best features of both local control and regionalism.

This dream is so politically implausible that it may not even be worth discussing, and it may make sense to focus only on more modest approaches. Perhaps it makes sense to consider some regional subsidies to education that are connected to the person, rather than the place, and that can be used to help pay for private education. Alternatively, there could be a role for regional policies that make it easier for people to go to school in different jurisdictions. This is a difficult area where the case for some degree of regional action seems merited, but it's hard to determine a course that will be remotely feasible politically and helpful.

TRANSPORTATION

Transportation is at the other extreme from economic development. The essence of transportation is in connecting different places. No one jurisdiction has the right incentives to build connections with neighboring jurisdictions.

After all, those other jurisdictions will receive benefits as well and unless those benefits are internalized, too little infrastructure will be built. Since the earliest days of the republic, there has been some recognition of the need for national investment in infrastructure. The Eisenhower highway system is, of course, the most classic example of national investment in car-based infrastructure.

Still, there are many reasons to think that the national investment in infrastructure is an imperfect approach. Over the past 20 years, federal funding for new projects has decreased. The federal government may also be too remote from local knowledge to understand the correct forms of infrastructure investment. A regional approach to ground-based transportation may make particular sense, because the regions have more information and incentives to get things right than the federal government, but a greater ability to internalize cross-jurisdiction externalities than local governments.

While the case for regionalism in transport planning and investment is clear, it is less clear what form that transport investment should take and I will not embarrass myself by going into it here. However, it is clear that different solutions will be appropriate for different regions. The ability to specialize investments is, after all, one of the big advantages of moving to a more regional approach. For example, high speed rail is much more likely to make sense in the Northeast than in the Arizona Sun Corridor.

One particular hope that I have is that regional transportation might lead to greater adoption of congestion charging on crowded roads. For 40 years, economists have argued that drivers should be charged for the costs that they impose on those around them. Today, we have ever-improving transponder technology that would make congestion charging easy to implement. There are particular gains to congestion charges that differ by time of day.

However, the advantages of congestion charging are reduced substantially if cars don't have the right electronic technology. If this technology differs from place to place, the costs of operation will be much higher. A common regional commitment to congestion charging and a region-wide system that uses the same transponder system and that follows the same rules will reduce the operational costs of the system significantly.

HOUSING

As in the case of education, there are advantages and disadvantages to regionalism in the case of housing policy. The case for local control is that there is a huge amount of local knowledge about both natural conditions and preferences that can be used by local planners. We have a long tradition of local control over planning and moving to a radically new system would be both jarring and wasteful of existing planning infrastructure.

On the other hand, there are many reasons to think that the current system is problematic both on empirical and theoretical grounds. As I discussed above, when bedroom suburbs make their own decisions, then they tend to take into account only the interests of neighboring homeowners who are generally opposed to new building. Larger jurisdictions tend to include employers and builders who have opposing interests. The empirical track record seems to suggest that larger jurisdictions tend to be less opposed to new development, in part because they allow for representation of a wider range of interest groups.

It is certainly true that when a jurisdiction makes decisions about new construction, it imposes externalities on other areas. For example, if an area pushes prices up by restricting supply, the people who currently don't live in the area but would like to are hurt by this increase in price. Employers are hurt by the need to pay more to their employees. There may even be transportation related externalities if lack of construction in an area close to an employment center pushes new units further out. The extra driving through areas between the new area and the area close to the employment center imposes traffic costs on the towns in between.

One interesting externality occurs because new construction in one community pushes new construction elsewhere. In a perfect world, where all communities imposed taxes or regulations to perfectly correct for externalities, this would not be a problem. However, if some communities do not impose barriers to new construction, then the barriers to construction in one place will inevitably lead to too much construction in another area. As I discussed above, it may be that new construc-

tion in the eastern towns of Massachusetts do create environmental costs, but if banning that construction leads to more construction in the Arizona desert, then this may actually increase the costs on the environment.

Historically, American regional growth has not been driven by differences in housing supply. Places grew because they were more economically productive or more attractive, not because zoning authorities were more permissive towards new housing. I am quite unsure whether this change is to the good, because I have seen little evidence that suggests to me that these regulatory authorities are appropriately weighing the costs and benefits of new construction. I am sure that some areas are currently too stringent and I suspect that authors may be too lenient.

A regional approach to land use regulation could improve the system. There are two natural ways in which regionalism might help. In some cases, regional land use planning could reduce the costs on localities of maintaining their own land use systems. There might be some chance of reducing bureaucratic duplication with some central provision of basic land use services.

The bigger advantage from regionalism lies in the possibility of pushing localities to better internalize the costs of their land use decisions. If I am right, and communities are making decisions that impose costs on outsiders, then regional land use planning could be used to prod local communities towards making better decisions. I suspect that this will require more than moral suasion. One approach would be to eliminate local control over land use alto-gether, but that seems as unlikely as moving to a voucher system in education. Moreover, it is true that localities have a great deal of local knowledge that would be lost by moving to decision making on a larger scale.

An intermediate solution requires the region to provide carrots and sticks to induce com-munities to build. If it is thought that too many high-price communities are permitting too little, then a point system could easily be put in place that would prod communities to build more. This point system would allocate quotas for new production on the basis of current density, prices, and location. High-cost areas would have a higher quota since there is more demand for their area. High-density areas would have a lower quota since they presum-ably have less land on which to build. Areas that are more proximate to employment centers could have a higher quota.

After these quotas are established, jurisdic-tions would gain points to meet their quotas on the basis of how much they build. The simplest system would just allocate one point for each new housing unit. A slightly more complex system would offer higher points for homes that were socially desirable, such as housing that was accessible to public transpor-tation or that was smaller and therefore more affordable to lower-income people. Com-munities would gain points by building and the quotas would be subtracted from the number of points earned.

This net point total would then be connected with cash payments to the communities. One system could feature only new aid to the com-munities that would be allocated on the basis of the number of points earned. This system would require more aid to jurisdictions, but it would have the benefit of facing less opposi-tion since no communities would lose aid relative to their current position.

A second system might be revenue neutral and require the jurisdictions that ended up below their quotas to take points from the jurisdictions that are above their quotas. The New Jersey system that has evolved since the Mount Laurel decision is, in a sense, a proto-type for a tradable points system of this kind. The goal is to ensure that communities bear the costs of their decisions not to build and such a points system could make that happen.

Of course, to be effective in creating new housing, the system would have to create strong incentives. In greater Boston, the mar-ginal benefit of building a unit to a community would have to be greater than $20,000 for it to have any effect. I suspect the figures in Northern California would have to be even greater. The exact details of any system should surely differ from region to region, but the case for some sort of regional planning in housing seems strong.

VI. CONCLUSION

In this paper, I have looked at the rise of regional economies and the implications of this rise. One hundred years ago, urban areas were dense centers surrounded by farmland. Together, population and employment spreads from place to place. This connection has created greater regional entities that differ greatly, but share the common attribute of relatively continuous density.

The 10 megaregions differ substantially in their incomes and economic productivity and they also differ in their growth rates. The older, denser regions have been declining. Growth is most dramatic in the lower density areas of the Southwest and Southern Florida. These growth patterns reflect a national trend where population growth is centered not in the most attractive or productive places, but in those places that don't constrain new construction by heavy land use regulations.

In the last section of the paper, I argued that the rise of regional economies increases the value for regional coordination in some areas, but not in others. Economic development policy seems to make more sense to continue as a local matter. Transportation planning, for example, is an obvious area where regional considerations are vital. Housing, on the other hand, increasingly needs a regional perspective since localities' decisions to block new construction is increasingly imposing costs on neighboring communities. I suggested a tradable points system that might encourage communities to internalize those costs while still respecting long-standing traditions of local control over development.

References

Beason, Richard and David Weinstein (1996). "Growth, Economies of Scale, and Targeting in Japan (1955–90)." *Review of Economics and Statistics* 78(2): 286–295.

Ciccone A, Hall R.E. (1996). "Productivity and the Density of Economic Activity." *American Economic Review* 86 (1): 54–70.

Glaeser, Edward L., Joseph Gyourko, and Raven Saks (2005). "Why is Manhattan So Expensive? Regulation and the Rise in House Prices." *Journal of Law and Economics* 48(2) (2005): 331–370.

Glaeser, E. and J. Gyourko (2003). "The Impact of Building Restrictions on Housing Affordability." *Federal Reserve Bank of New York Economic Policy Review* 9(2):21–39.

Glaeser, Edward L., Matthew Kahn, and Jordan Rappaport (2007). "Why Do the Poor Live in Cities?" *Journal of Urban Economics*, forthcoming.

Glaeser, Edward L. and Matthew Kahn (2004). "Sprawl and Urban Growth." In *The Handbook of Regional and Urban Economics*, Volume 4 (V. Henderson and J. Thisse, eds.), North Holland Press.

Glaeser, Edward L. and Matthew Kahn (2001). "Decentralized Employment and the Transformation of the American City." *Brookings-Wharton Papers on Urban Affairs* 2 (2001).

Glaeser, Edward L. and Janet E. Kohlhase (2004). "Cities, Regions, and the Decline of Transport Costs." *Papers in Regional Science* 83(1): 197–228.

Glaeser, Edward L., Jed Kolko, and Albert Saiz (2001). "Consumer City" (joint with J. Kolko and A. Saiz). *Journal of Economic Geography* 1: 27–50.

Glaeser, Edward L. and David Mare (2001). "Cities and Skills." *Journal of Labor Economics* 19(2) (2001): 316–342.

Glaeser, Edward L. and Albert Saiz (2004). "The Rise of the Skilled City" (joint with A. Saiz). *Brookings-Wharton Papers on Urban Affairs* 5: 47–94.

Glaeser, Edward L. and Jesse Shapiro (2003). "Urban Growth in the 1990s: Is City Living Back?" (joint with J. Shapiro). *Journal of Regional Science* 43(1): 139–165.

Glaeser, Edward L. and Bryce Ward (2006). "The Causes and Consequences of Land Use Regulation: Evidence from Greater Boston." NBER Working Paper #12601.

Greenstone, Michael and Enrico Moretti (2003). "Bidding for Industrial Plants: Does Winning a 'Million Dollar Plant' Increase Welfare?" NBER Working Paper #9844.

Gyourko, Joseph and Albert Saiz (2004). "Reinvestment in the Housing Stock: The Role of Construction Costs and the Supply Side." *Journal of Urban Economics* 55(2):238–256.

Holmes, Thomas J. (1998). "The Effects of State Policies on the Location of Industry: Evidence from State Borders." *Journal of Political Economy* 106 (4): 667–705.

Hoxby, Caroline M. (2001). "All School Finance Equalizations Are Not Created Equal." *Quarterly Journal of Economics* 116 (4).

Hoxby, Caroline M. (2000). "Does Competition Among Public Schools Benefit Students and Taxpayers?" *American Economic Review,* 90 (5).

Katz, L. and K. T. Rosen (1987). "The Interjursdictional Effects of Growth Controls on Housing Prices." *Journal of Law and Economics* 30(1): 149–160.

Jakabovics, A. (2006). "Housing Affordability Initiative: Land Use Research Findings." http://web.mit.edu/cre/research/hai/land-use.html.

Logan, John and Harvey Molotch (1988). *Urban Fortunes.* Berkeley: University of California Press.

Margo, Robert (1992). "Explaining the Postwar Suburbanization of the Population in the United States; the Role of Income." *Journal of Urban Economics* 31: 301–310.

Moretti, Enrico (2004). "Estimating the Social Return to Higher Education: Evidence from Longitudinal and Repeated Cross-Sectional Data." *Journal of Econometrics* 121 (1–2).

Rauch, James E. (1993). "Productivity Gains from Geographic Concentration of Human Capital: Evidence from the Cities." *Journal of Urban Economics* 34 (3): 380–400.

Regional Plan Association (2006). "America 2050: A Prospectus." New York.

* I am enormously grateful to Nina Tobio who provided splendid research assistance on this paper. The Taubman Center for State and Local Government provided research funding.

Megaregions: Benefits beyond Sharing Trains and Parking Lots?

Saskia Sassen
University of Chicago

INTRODUCTION

I have been asked to examine whether there are particular advantages to economic interactions at the megaregional scale and whether such interactions might play a role in enhancing the advantages of megaregions in today's global economy. Familiar advantages of scales larger than that of the city, such as metropolitan and regional scales, include the benefits of sharing transport infrastructures for people and goods, enabling robust housing markets, and, possibly, supporting the development of office, science, and technology parks. Critical policy options identified by the Regional Plan Association (RPA) in this regard would aim at strengthening the megaregional scale of economic interactions by investing in intercity and high-speed regional rail, enhanced goods movement systems, and land-use planning decisions.

More complex and elusive is whether the benefits of megaregional economic interaction can go beyond these familiar scale economies and, further, whether this would strengthen the position of such megaregions in the global economy. It is this set of issues that I have been asked to address. Much of the effort involves developing an analytic framework that allows us to begin to understand the parameters of these two issues. There is no definitive research on this subject. Thus empirical specification can only be partial as the available evidence is fragmentary for the regional level, a shortcoming that becomes acute when dealing with the novel category of the megaregion.[1] There is, however, enough analysis and evidence on one particular component of this subject—the advantages for global firms and markets of particular types of agglomeration economies at the urban level—that we can begin to use it as a lens onto the megaregional scale. Agglomeration economies are to be distinguished from familiar urbanization economies (the advantages of scale and spatial concentration). They involve complex interactions of diverse components, not simply, for instance, more people using a train line and the scale economies this might enable.

If we are to understand the advantages of economic interaction at the megaregional scale we need to go beyond explaining why we have megaregions today. I do not want to diminish the critical importance of understanding the characteristics of the 10 megaregions and their "urbanization" advantages.[2] It is critical to understand the causes or combinations of dynamics that produce the 10 megaregions identified by RPA. However, that is not enough.

Megaregions might, after all, be merely the result of population growth in a geographic setting where cities and metro-areas blend into each other. Voila, a megaregion! And this does indeed call for cross-regional infrastructures, notably transport and electricity, and some of the other components of regional planning and coordination presented in the RPA report. But are these conditions, which amount to an expanded version of urbanization economies, enough to answer the larger questions raised by RPA and the Princeton Policy Institute for the Region? My answer is a qualified no.

SPECIFYING MEGAREGIONAL ADVANTAGE BEYOND SCALE ECONOMIES

One central argument I develop in this paper is that the specific advantages of the megaregional scale consist of and arise from the coexistence within one regional space of multiple types of agglomeration economies. These types of agglomeration economies today are distributed across diverse economic spaces and geographic scales: central business districts, office parks, science parks, transportation networks, housing markets, low-cost manufacturing districts (today often offshore), tourism destinations, specialized branches of agriculture, such as horticulture or organically grown food, and the complex kinds of agglomeration economies evident in global cities. Each of these spaces evinces distinct agglomeration economies and empirically at least, is found in diverse types of geographic settings—from urban to rural, from local to global.

The thesis is that a megaregion is sufficiently large and diverse so as to accommodate a far broader range of types of agglomeration economies and geographic settings than it typically does today. This would take the advantages of megaregional location beyond the notion of urbanization economies. A megaregion can then be seen as a scale that can benefit from the fact that our complex economies need diverse types of agglomeration economies and geographic settings, from extremely high agglomeration economies evinced by the specialized advanced corporate services to the fairly modest economies evinced by suburban office parks and regional labor-intensive low-wage manufacturing. It can incorporate this diversity into a single economic megazone. Indeed, in principle, it could create conditions for the return of particular (not all) activities now outsourced to other regions or to foreign locations.[3]

Thus the critical dimension for the purposes of this paper is not just a question of the contents of a megaregion, such as its economic sectors, transport infrastructure, housing markets, and types of goods and services that get produced, distributed, exported, and imported. Equally critical is the specification of economic interactions within the megaregion in order to detect what could be reincorporated into that region (e.g., factories or routine clerical work that is now outsourced to other national or foreign areas) as well as to detect emerging megaregional advantages. All of this requires working off the fact that these megaregions exist and the specific indicators used by RPA in identifying those regions. But it also requires going beyond these indicators and the realities they are meant to represent.

One way of specifying some of this empirically is to establish *whether agglomeration economies (not just urbanization economies) matter for developing the spatial organization of a megaregion.* Examining the question of agglomeration economies in the current period is framed by two facts that are potentially in tension with each other. On the one hand, the new information technologies enable firms to disperse a growing range of their operations, whether at the metro, regional, or global level, without losing system integration;[4] this has the potential to reduce(though not eliminate) the benefits of urbanization economies for such firms. On the other hand, the evidence clearly shows the urbanizing and increasing density of massive regions, including scale-ups to the megaregional level as identified by RPA.[5]

I first address the most extreme instance—globalized firms with considerable digitization of their production process and their outputs. In this case there are conceivably fewer and fewer agglomeration advantages, especially for the most advanced sectors, typically high-value producing, able to buy the latest technologies, and highly globalized, that is, with multiple operations across the world. However, in contrast to this technologically driven explanation, I describe how and why precisely these firms are subject to extreme agglomeration economies in some—not all—of their components.[6] This fact matters for understanding the potential advantages of the megaregion, which certainly do contain extremely dense cities with diverse resources and types of talent.

Furthermore, while the "multi-sited" character of the leading economic sectors includes cit-

ies as one key site, these advanced sectors often require other types of locations—some marked by medium and even low or no agglomeration economies, with some economies of scale, but strong preferences for low-cost, often underdeveloped areas. This combination of dense cities with less dense development starts to describe what might be a megaregion.

And once we have started to describe a new economic structure that is ideally suited to the megaregional scale, we need to consider how it could evolve over time. So, we need to examine existing and new types of economic growth at that megaregional scale. For example, if a megaregion is able to capture all or most of the components of the production chains or value chains of "multi-sited" firms, long-term growth may exceed what we see in a pre-megaregional economy. Elsewhere, I have discussed how a firm's central headquarter functions expand as a result of a firm's geograpahic dispersal (whether national or global) of its activities—factories, back offices, service outlets, affiliates, etc.[7] Thus the co-presence within a megaregion of a firm's top-level headquarters and low-cost routine work could add yet another specific source of growth for that megaregion in both its cities and its low-cost underdeveloped areas. That is to say, this is a growth effect that builds on but is distinct from the mere addition of jobs resulting from that megaregion capturing more sites of a firm's chain of operations.

Now the question becomes: *Can a megaregion seek to accommodate a larger range of the operations constituting a firm's value chain—from those subject to agglomeration economies to those that do not evince such economies?*

Practically speaking, this points to the possibility of bringing into (in some cases, back to) a megaregion some of the services and goods now produced offshore to get at lower wages and less regulations. Can these be reinserted in the low-growth, low-cost areas of a megaregion? What type of planning would it take, and can it be done so as to optimize the benefits for all involved, firms, workers, localities? This would expand the project of optimizing growth beyond the usual suspects—office and science parks being one notable example—and move across far more and more diverse economic sectors and types of local economies within the megaregion. It would use the lever of the megaregional scale to provide diverse spaces catering to different types of activities, ranging from those subject to high to those subject to low agglomeration economies. And, finally, the megaregional scale would help in optimizing the growth effect arising from the interactions of some of these different kinds of agglomeration economies. This growth effect would be optimized by re-regionalizing some of the low-cost operations of firms headquarted in that megaregion but with many of its lower value-added operations spread across the country and/or the world.

If this type of thesis does indeed capture a potential of megaregions, it would be the making of new economic history. The possibility of this type of potential is easily obscured by the prevalence of national level economic indicators, data sets, and policies. Identifying the megaregion produces a new, intermediate level,

one that even though partly dependent on national macro-policies also inserts a far more specific set of issues into the economic picture.[8] A megaregion can combine a very large share of the diverse economies that are very much part of our current era. And it can incorporate growth effects arising from the interactions of some of these diverse economies.

This way of thinking about the megaregional scale raises the importance of planning and coordination to secure optimal outcomes for all parties involved, including the challenge of securing the benefits firms are after when they disperse their operations to low wage areas. This would work for some types of economic sectors and types of firms, not for all. We know that some activities that have been outsourced to other countries have not worked out and have been repatriated—they range from airline sales agents to particular types of design work in industries as diverse as garments and high-tech. But many of these outsourced activities are doing fine as far as the firms are concerned. We need research and specific policies to establish the what, how, and where of the advantages for the pertinent firms of accessing low-wage workers in the U.S.; this includes understanding how location of these low-cost components in the megaregion where a given firm is headquartered could compensate for higher costs. And it would mean ensuring that such repatriated low-wage jobs become building blocks for development and viable livelihoods. This may require megaregional investment in developing low-cost areas for such jobs—a kind of rural enterprise zone.

There is possibly a positive macro-level effect from repatriating some of these jobs if a race to the bottom can be avoided and a certain level of consumption capacity secured via reasonable wages or particular indirect subsidies. This brings a specific positive effect for a megaregion's less developed areas insofar as lower-wage households tend to spend a much larger share of their income in their place of residence—they lack the investment capital of upper-income strata who can wind up allocating most of their income on overseas investments. Finally, this is also one element in the larger challenge of securing more equitable outcomes.[9] Since there is sufficient evidence about how extreme maldistribution of the benefits of economic growth is not desirable in the long run, we need to ask about the distributive effects of the current configuration and of (potentially) optimized outcomes as described in this paper.

These ways of specifying the meaning of a megaregion (or a region) take us from a "packaging" approach to a more dynamic concept of the megaregion: beyond urbanization advantages, a megaregion may well turn out to be a sufficiently large scale to optimize the benefits of containing multiple and interacting local economies.

PROXIMITY AND ITS ADVANTAGES: DOES IT HOLD FOR MEGAREGIONS?

Today's information technologies and communication capabilities can deliver system integration no matter how far-flung the operations of a firm or sector might be. If all firms and sectors can buy/use these technologies to reduce or neutralize agglomeration economies/advantages, the result would be a decline in the benefits of locations that deliver agglomeration economies, most notably global cities. Such a decline would be further strengthened by the possibility of rising shares of e-commuters—working online from home.

In its most extreme version, this scenario suggests that the advantages of locating in a megaregion would be limited to urbanization economies. Firms need to locate somewhere and so do their workers, so why not a megaregion; and, secondly, regardless of whether there are specific megaregional locational advantages, there would be a demand for local suppliers of final and intermediate goods and services that need to be produced *in situ*—that cannot be imported from far away, or at least not yet. The fact itself of population growth—a fact in most of the RPA-defined megaregions—is enough to feed this type of demand.

Under these conditions, the specificity of megaregional locational advantages comes down to the fact that there is a market, or rather a whole range of markets for needed goods and services, both final and intermediate. Transport, housing, office buildings, factory buildings, and so on, all meet a real demand by households, governments, and their multiple instances, from schools to courts, institutions of all sorts, and firms. As populations and distances grow, novel types of demand emerge: for speed-rail, super highways, more diversity in the housing supply. No matter how complex the components of this final and intermediate demand, this is an elementary version of the advantages of the megaregional scale.

But, are there more complex advantages for megaregional location?

The starting point is that location is a variable. The firm that can replace agglomeration advantages with the new information technologies represents one extreme case on the location variable: it evinces minor if any agglomeration economies. The fact of population growth, and the associated need for housing and all that comes with it, is in many ways the same type of point on that variable; the difference is that it is subject to urbanization advantages. A very different case is high agglomeration economies; this is well established for very specialized branches of global finance and the most innovative branches of high-tech industries, with global cities and silicon valleys the respective emblematic spatial forms.

The advantages of location in a megaregion for these diverse types of cases need to be empirically specified. We can establish that in the first two cases above the particular advantage is some very broad, and geographically expanded, notion of urbanization advantages—the bundle of infrastructures, labor markets, buildings, housing, basic institutional resources, amenities. In a megaregion these advantages spread over a vast geographic terrain, engendering its own specific components of final and intermediate demand, e.g., rapid-transit systems.

The question then becomes how do we enhance these urbanization advantages, how do we avoid excess growth/expansion/spread and its negative effects on congestion, prices, costs, etc.? Whether markets or planning are

the desirable instruments to optimize "urbanization" economies (broadly understood in that they include not only urban locations) will depend on a range of variables. One potentially innovative line of analysis here is the extent to which the megaregion enables novel ways of handling negative externalities: what is one sector's or location's negative externality, is another's potential business opportunity, a point I return to later.

On the other hand, in the case of sectors subject to agglomeration economies, it may well be the case that the megaregion does not contain distinctive advantages over other scales, notably cities and metro areas. What these sectors seem to need is a bundle of resources that correlate with high-density, and, at its extreme, very dense central places—such as global cities and silicon valleys. The question then becomes whether there is one or several specific types of agglomeration economies that can develop, and be enhanced, at the scale of the megaregion. The megaregions identified by RPA all contain high-density locations; a firm subject to agglomeration economies may well find the mix of highly specialized diverse resources it needs in one of those locations. But does it need a whole megaregion attached to that location?

Here we enter new theoretical and empirical territory. One critical hypothesis I developed for my global city model is that insofar as the geographic dispersal of a global firm's operations (whether factories, offices, or service outlets) feeds the complexity of central headquarter's work, the more globalized a firm the higher the advantages its headquarters derive from central locations that concentrate multiple resources and talents (see footnote 6).[10] If I were to adapt this to the megaregion, one inference is that the advantage of a megaregional scale is that it could, in principle, contain both the central headquarters and at least some of those dispersed operations of global firms. In other words, is a megaregion a scale at which such firms can actually also "outsource jobs" and suburbanize back office work—both in search of cheaper costs—and benefit from the region's major city(s), including in some cases, global cities (New York City, Chicago, Los Angeles, Boston, San Francisco), or cities with significant global-city functions (Minneapolis, Miami, Atlanta, among others) for their headquarter functions.

Can megaregions deliver particular advantages if they can also contain some of the geographically dispersed operations of a firm?

The evidence shows that increasingly the spatial organization of firms and economic sectors contains both points of spatial concentration and points of dispersal. Further, the evidence also shows that in many cases these points of spatial concentration contain segments in a firm's chain of operations that evince rather strong agglomeration economies. One underlying (and disciplining) trend here, becoming visible already in the 1970s, is that spatial concentration is costlier for many firms so that the push is to disperse whatever operations can be dispersed; this contrasts with earlier periods when even large headquarters kept all functions in one place. This dispersal of a firm's operations can be at a regional, national and/or global level, and sites that deliver

agglomeration economies can vary sharply in content and in the specifics of the corresponding spatial form.[11] For instance, Chicago's financial center, Los Angeles's Hollywood, Northern California's Silicon Valley, each deliver (distinctive) agglomeration economies to firms and sectors which also contain often vast geographic dispersal of some of their other operations.[12]

A focus on the fact that much economic activity contains both spatial concentration and trans-local chains of operations helps us situate the specifics of a city, a metro area, or a megaregion in a far broader systemic condition, one that might include both points subject to sharp agglomeration economies and points that are not—where geographic dispersal is an advantage. What the megaregion offers in this context is a bigger range of types of locations than a city or a metro area—from locations subject to high agglomeration economies all the way to locations where the advantage comes from dispersal.

Taking it a step further, in my own research I found that the most globalized and innovative firms were characterized by the fact that agglomeration economies are themselves partly a function of dispersal. That is to say, the more globalized and thus geographically dispersed a firm's operations, the more likely the presence of agglomeration economies in particular moments (top-level headquarter functions) of that firm's chain of operations.[13] For the purposes of this essay, it underlines the fact of a single dynamic with diverse spatial manifestations, i.e. both agglomeration and dispersal, across diverse geographic scalings;

a megaregion would then conceivably be a scaling that not only can incorporate these different settings, but also can benefit from the growth effects of their interaction, a point I develop next.

One way of specifying some of this empirically is to posit a direct interaction effect between growth in respectively a megaregion's locations for dispersed economic activities and locations for activities subject to high agglomeration economies. The more the former grow, the more the latter will also grow. The trick is then to maximize the co-presence in a given megaregion of these two types of locations. It is important to notice that this also sets limits to the advantages of urbanization economies insofar as these would raise the costs of operating in locations marked by low-density (dispersed operations). Urbanization economies turn out to be defined by a curve: they grow with scale, but up to a point. That point is typically specified in terms of negative externalities. But what my analysis here suggests is that this point can also be specified in terms of the economic losses derived from not allowing the "development" of dispersal locations; since this means locations where firms can send their low-wage jobs requiring little education, it clearly goes against the prevailing aims of most places, which is to get high-wage, high-capital intensive jobs. Critical here is to recognize that even the most advanced economic sectors have these types of jobs. Finally, if what is today the point on the curve where familiar negative externalities set in (e.g., excess congestion) can be made to coincide with that development of "dispersal locations" for the low-wage component of

firms, we would be making an advantage out of what is now a disadvantage. Again, critical for me is the fact that by keeping such jobs in the U.S., we can avoid a race to the bottom and develop disadvantaged localities.

In practical terms there are, clearly, massive challenges for a megaregion to achieve this type of co-presence—maximizing the extent to which a megaregion can contain both the agglomeration and dispersal segments of a firm's chain of operations. For one, it is a countersensical, counterintuitive proposition. It is difficult to see why a megaregion's highly dynamic economic spaces (the central areas of its global cities and silicon valleys), anchored by the headquarters of global and national firms, might actually be partly fed and strengthened by developing the "dispersal locations" of those same firms. Thinking of developing such "dispersal locations" as one way of making the most of negative externalities might make it more acceptable to the skeptics—you might as well go for activities that benefit from geographically dispersed arrangements once you hit excess congestion disadvantages. But one option at this point is of course such items as golf courses and ex-urban luxury housing. This is an argument that could be countered since the megaregions identified by the RPA contain much land that is not optimal for such uses, but that could be optimal for developing "dispersal locations." Importantly, some areas that are currently disadvantaged could benefit from such development, and a race to the bottom is avoided.

The megaregion can then be seen as an interesting scalar geography: it can contain some of the dispersals of a firm's operations that feed these new kinds of agglomeration economies. It would suggest that strategic regional planning could aim at maximizing the combination of different locational logics. It is this combination that in my view marks the specificity of the "project" contained in the notion of the megaregion. This kind of region cannot be looked at simply as an outcome: there it is, and let's then find a packaging that brings a lot of this together under one umbrella. As a term, megaregion has a certain passivity attached to it. "Megaregional agglomeration economies," on the other hand, is a notion that captures a dynamic that produces outcomes. This in turn opens up a research agenda: for instance, to understand at what territorial scales such economies are enhanced or become weaker. Megaregion is, however, a catchy term, describing a self-evident condition, and in that sense is an acceptable and digestible term (something we cannot say about "megaregional agglomeration economies").

My hypothesis here could be framed as follows: The more an urban region is being shaped by the new economic dynamics, the more its spatial organization will involve agglomeration economies. These particular agglomeration economies are a function of the added management complexities for firms that disperse their economic activities under conditions of systemic integration, no matter the scale—regional, national, or global. The challenge for a megaregion is to incorporate economic both headquarter functions and those dispersed activities. In brief, megaregions should aim at maximizing the incorporation of diverse spatial logics.

The next section examines one critical aspect of such co-presence: does geographic dispersal feed agglomeration economies? I take the extreme case—the most digitized and globalized firms—as a natural experiment to understand the parameters of the articulation between geographic dispersal and agglomeration economies, and what it would mean to regionalize this articulation.

DOES GEOGRAPHIC DISPERSAL FEED AGGLOMERATION ECONOMIES?

A good starting point is to focus on why the most advanced firms of the knowledge economy are subject to what seem often extreme agglomeration economies, even when they function in electronic markets and produce digitized outputs. Another way to ask it is by focusing on the most globalized and digitized of all knowledge sectors: Why does global finance need financial centers? Or, more generally, why do highly specialized global corporate services that can be transmitted digitally thrive in dense downtowns? This means inserting place in an analysis of knowledge economies that are usually examined in terms of their mobility and space-time compression. Looking at the knowledge economy, and more broadly, global firms, from the optic of regions, cities, or metro-areas, brings in different variables.[14]

Much is known about the wealth and power of today's global firms. Their ascendance in a globalizing world is no longer surprising. Similarly, with the new information and communication technologies, much attention has focused on their enormous capacities for worldwide operations without losing central control. Less clear is why cities or regions should matter for global firms; particularly, global firms that are rich enough to buy whatever the technical innovations that free them from place, its frictions, and its costs.

There are several logics that explain why cities matter to the most globalized (dispersed) and digitized firms and sectors in a way they did not as recently as the 1970s. Here I briefly focus on three of these logics.[15]

The first one is that no matter how intensive a user of digital technology a firm is, its operational logic is not the same as the engineer's logic for designing that technology. Confusing these two potentially very diverse logics has produced a whole series of misunderstandings. When the new information and communications technologies (ICTs) began to be widely used in the 1980s, many experts "forecasted" the end of cities as strategic spaces for firms in advanced sectors. Many routinized sectors did leave cities and many firms dispersed their more routine operations to the regional, national, and global scale. But the most advanced sectors and firms kept expanding their top-level operations in particular types of cities.

Why were those experts so wrong? They overlooked a key factor: when firms and markets use these new technologies they do so with financial or economic objectives in mind, not the objectives of the engineer who designed the technology. The logics of users may well thwart or reduce the full technical capacities of the technology.[16] When firms and markets disperse many of their operations

globally with the help of the new technologies, the intention is not to relinquish control over these operations. The intention is to keep control over top-level matters and to be capable of appropriating the benefits/profits of that dispersal.[17] Insofar as central control is part of the globalizing of activities, their top-level headquarter functions actually have expanded because it is simply more complicated and riskier to function in 30 or 50 or more countries, each with distinct laws, accounting rules, and business cultures.

As these technologies are increasingly helpful in maintaining centralized control over globally dispersed operations, their use has also fed the expansion of central operations. The result has been an increase in high-level office operations in major cities and a growth in the demand for high-level and highly paid professional services, either produced in-house or bought from specialized service firms. Thus the more these technologies enable global geographic dispersal of corporate activities, the more they produce density and centrality at the other end—the cities where their headquarter functions get done.

A second logic explaining the ongoing advantages of spatial agglomeration has to do precisely with the complexity and specialization level of central functions. These rise with globalization and with the added speed that the new ICTs allow. As a result global firms increasingly need to buy the most specialized financial, legal, accounting, consulting, and other such services. These service firms get to do some of the most difficult and speculative work. It is increasingly these corporate service firms that evince agglomeration economies. Their work benefits from being in complex environments that function as knowledge centers, which contain multiple other specialized firms and high-level professionals with worldwide experience. Cities are such environments—with the 40-plus global cities in the world the most significant of these environments, but a growing number of other cities strong in particular elements of such environments. In brief, cities or central places provide the social connectivity which allows a firm to maximize the benefits of its technological connectivity.[18]

A third logic concerns the meaning of information in an information economy. There are two types of information. One is the datum, which may be complex yet is standard knowledge: the level at which a stock market closes, a privatization of a public utility, a bankruptcy. But, there is a far more difficult type of "information," akin to an interpretation/evaluation/judgment. It entails negotiating a series of datums and a series of interpretations of a mix of datums in the hope of producing a higher-order datum. Access to the first kind of information is now global and immediate (even if often for a high fee) from just about any place in the highly developed world and increasingly in the rest of the world thanks to the digital revolution.

But it is the second type of information that requires a complicated mixture of elements —the "social infrastructure" for global connectivity—which gives major financial centers a leading edge. When the more complex forms of information needed to execute major international deals cannot be acquired from existing databases, no matter what one can pay, then one needs to make that information; it becomes part of the production process in specialized corporate service firms, including financial services both as service providers and as firms in their own right. That making includes as a critical component interpretation, inference, and speculation. At this point one needs the social information loop and the associated de facto interpretations and inferences that come with bouncing off information among talented, informed people. It is the importance of this input that has given a whole new importance to credit rating agencies, for instance. Part of the rating has to do with interpreting and inferring. When this interpreting becomes "authoritative" it becomes "information" available to all. For specialized firms in these complex domains, credit ratings are but one of these inputs; the making of authoritative information needs to be part of a production process, either in-house or bought from specialized firms. This process of making inferences/interpretations into "information" takes an exceptional mix of talents and resources. Cities are complex environments that can deliver this mix.

The key implication of this analysis for megaregions is the possibility of containing both (at least some of) the dispersed operations of a given firm and the central headquarter operations. The feedback effects of containing both can be significant, feeding simultaneously growth in a megaregion's low-cost possibly marginal areas and in its global cities, or cities that are national business centers.

THE ONGOING IMPORTANCE OF CENTRAL PLACES

Cities have historically provided national economies, polities, and societies with something we can think of as centrality. The usual urban form for centrality has been density, specifically the dense downtown. The economic functions delivered through urban density in cities have varied across time. But it is always a variety of agglomeration economies, no matter how much their content might vary depending on the sector involved. While the financial sector is quite different from the cultural sector, both evince agglomeration economies; but the content of these benefits can vary sharply. One of the advantages of central urban density is that it has historically helped solve the risk of insufficient variety. It brings with it diverse labor markets, diverse networks of firms and colleagues, massive concentrations of diverse types of information on the latest developments, diverse marketplaces. The new information and communication technologies (ICTs) should have neutralized the advantages of centrality and density. No matter where a firm or professional is, there should be access to many of the needed resources. But in fact, the new ICTs have not quite eliminated the advantages of centrality and density, and hence the distinct role of cities for leading global firms.[19]

Even as much economic activity has dispersed, the centers of a growing number of cities have expanded physically, at times simply spreading and at times in a multi-nodal fashion. The outcome is a new type of space of centrality in these cities and their metro-areas: it has physically expanded over the last two decades, a fact we can actually measure, and it can assume more varied formats. The geographic terrain for these new centralities is not always simply that of the downtown; it can be metropolitan and even regional. In this process, the geographic space in a city or metro area that becomes centralized often grows denser as measured in number of firms, though not necessarily households, than it was in the 1960s and 1970s. This holds for cities as different as Zurich and Sydney, Sao Paulo and London, Shanghai and Buenos Aires. (But population density is not necessarily the best indicator of this type of density.)

The global trend of expanded newly built and rebuilt centralized space suggests an ironic turn of events for the impact of ITCs on urban centrality. Clearly, the spatial dispersal of economic activities and workers at the metropolitan, national, and global level that began to accelerate in the 1980s actually is only half the story of what is happening. New forms of territorial centralization of top-level management and control operations have appeared alongside these well-documented spatial dispersals. National and global markets as well as globally integrated operations require central places where the work of globalization gets done, as analyzed in the preceding section.

Centrality remains a key feature of today's global economy. But today there is no longer a simple straightforward relation between centrality and such geographic entities as the downtown, or the central business district (CBD). In the past, and up to quite recently in fact, centrality was synonymous with the downtown or the CBD. Today, partly as a result of the new ICTs, the spatial correlates of the "center" can assume several geographic forms, ranging from the CBD, the metro area, to the new global grid comprising global cities.[20]

Particular urban, metro, and regional spaces are becoming massive concentrations of new technical capabilities. A growing number of buildings are the sites for a multiplication of interactive technologies and distributed computing. And particular global communication infrastructures are connecting specific sets of buildings worldwide, producing a highly specialized interactive geography, with global firms willing to pay a high premium in order to be located in it. For instance, AT&T's global business network now connects about 485,000 buildings worldwide; this is a specific geography that actually fragments the cities where these buildings are in as you need to be in a "member" building to access the network. The most highly valued areas of global cities, particularly financial centers, now contain communication infrastructures that can be separated from the rest of the city, allowing continuous upgrading without having to spread it to the rest of the city. And they contain particular technical capabilities, such as frame relays, which most of the rest of the city does not. Multiplying this case for thousands of multinational firms begins to give us an idea of these new intercity connectivities, largely invisible to the average resident.

One question is whether some of these trans-local operations are actually located within some of the megaregions that concern us here? This is an empirical question, but one with policy/planning implications. Similarly, if we consider these globally networked spaces of centrality as platforms for global operations of firms and markets, we might ask what components of these platforms are contained within a given megaregion. Finally, these platforms consist of a variety of specific geographic sub-national spaces but also electronic spaces. We might then also ask what are the implications for megaregions of the fact that a growing number of sub-national scales—from cities to precisely such megaregions—emerge as strategic territories that contribute to articulate a new global political economy, and new national and regional political economies.

REGIONAL SPECIFICITY AND KNOWLEDGE ECONOMIES: ANY LINKS?

Let me start by saying that the answer is yes. How much a region's specificity matters will vary, partly depending on that region's economy. My point is that a region's specificity matters more than is usually assumed, and that it matters in ways that are not generally recognized. The policy implications of the argument I develop in this and the ensuing section is that we have focused far too much on competition—between cities, between regions, between countries—and not enough on the emergence of new types of networked systems and the partly associated emergence of increasingly specialized global divisions of functions. These networked systems arise partly out of two trends. The first is the increasing prevalence of multi-sitedness of firms. The second is the evolution of global markets into global platforms that are open to and from many different places. And the increasingly specialized global division of functions arises out of the multiplication of specialized economic sectors and the increasing complexity of many of these sectors.

Among the key implications for megaregions of these combined trends is that their scale can allow them to capture a large share of those networks, and, secondly, that the specialized economic strengths of a region increasingly matter. Yes, there is competition, but it accounts for far less than is usually assumed. What really matters is the specialized difference of a city or region. In this section I examine the connection between regional economic specificity and the formation of advanced knowledge economies. And in the ensuing section I examine the question of increasingly homogenized landscapes and built environments to understand how regional or urban specificity can coexist with that homogenizing.

How does a city or a region become a knowledge economy? Let me use the case of Chicago to illustrate what I am trying to get at here. It is common to see Chicago as a latecomer to the knowledge economy (and thus to global city status). Why did it happen so late—almost 15 years later than in New York and London? Typically the answer is that Chicago had to overcome its agro-industrial past, that its economic history put it at a disadvantage compared to old trading and financial centers such as New York and London.[21]

But I found that its past was not a disadvantage. It was one key source of its competitive advantage. The particular specialized corporate services that had to be developed to handle the needs of its agro-industrial regional economy gave Chicago a key component of its current specialized advantage in the global economy.[22] While this is most visible and familiar in the fact of its preeminence as a futures market built on pork bellies, so to speak, it also underlies other highly specialized components in its global city functions. The complexity, scale, and international character of Chicago's historical agro-industrial complex required highly specialized financial, accounting, and legal expertise, quite different from the expertise required to handle the sectors New York specialized in—service exports, finance on trade, and finance on finance. Today there are other sectors that are, clearly, also critical to Chicago's advanced service economy, notably the conventions and entertainment sector and cultural industries. But the point here is that Chicago's past as a massive agro-industrial complex gave the city some of its core and distinctive knowledge economy components.

But for this specialized advantage to materialize, past knowledge needed to be adapted into a different set of economic circuits. This required, then, disembedding that expertise from an agro-industrial economy and re-embedding it in a "knowledge" economy—that is to say, to an economy where expertise can increasingly be commodified, function as a key input, and, thereby constitute a new type of intermediate economy. Having a past as a major agro-industrial complex makes that switch more difficult than a past as a trading and financial center. This then also explains partly Chicago's "lateness" in bringing that switch about. But that switch is not simply a matter of overcoming that past. It requires a new organizing logic that can revalue the capabilities developed in an earlier era (Sassen 2006a: chapters 1 and 5).[23] It took quite an effort to execute the switch.

Through its particular type of past, Chicago illuminates aspects of the formation and the specifics of knowledge economies that are far less legible in cities such as New York and London, which even though they did have manufacturing were dominated by predominantly trading and banking economies. A first issue is then that Chicago's past as an agro-industrial economy points to the mistake of assuming that the characteristics of global cities correspond to those of such old trading and banking centers.

A second issue raised by the Chicago case is that while there are a number of global city regions today with heavy manufacturing origins, many once important manufacturing cities have not made the switch into a knowledge economy based on that older industrial past. Along with Chicago, Sao Paulo, Tokyo, Seoul, and Shanghai are perhaps among today's major global city regions with particularly strong histories in heavy manufacturing. But most once important manufacturing cities, notably Detroit and the English manufacturing cities, have not undergone that type of switch. They were to some extent dominated by a single or a few industries and shaped up more like monocultures. This points to the importance of thresholds in the scale and diversity of a region's manufacturing past to secure the components of knowledge production I identify in Chicago's case—specialized servicing capabilities that could be dislodged from the

organizational logic of heavy manufacturing and re-lodged in the organizational logic of today's so-called knowledge economy.

The specialized economic histories of major cities and regions matter in today's global economy because they are the main way in which national economies are inserted in variable ways in multiple globally networked divisions of functions. It never was "the" national economy that articulated a country with the international division of functions. But today it is even less so because the global economy consists of a vast number of particular circuits connecting particular components of cities and regions across borders. It is at this level of disaggregation that it is best to understand how cities and regions are globally articulated. It is also in this context that we can begin to see how much more the specialized economic histories of a region matter today than they did in the Keynesian period marked by national territorial convergence and mass production, rather than today's period marked by proliferation of increasingly specialized and diverse services.

Thus, returning to our example, Chicago today has a specialized advantage in producing certain types of financial, legal, and accounting instruments because financial, legal, and accounting experts in Chicago had to address in good part the needs of the agro-industrial complex; they had to deal with steel and with cattle produced for the regional, national, and international markets. It is this specialized type of knowledge that matters for Chicago's competitive situation in the global market. Chicago, Sao Paulo, Shanghai, Tokyo, and Seoul are among the leading producers of these types of specialized corporate services, not in spite of their economic past as major heavy industry centers, but because of it.

The fact that these distinctions and differences in the specialized economic histories of cities and regions become increasingly prominent and value-adding in today's global, and also national, economy is easily obscured by the common emphasis on competition and cross-border standardization. Competition and standardization have been rescaled partly to the subnational level of cities and regions—this is a reality that it is difficult to avoid. But the emphasis remains on competition, notably intercity competition, and on standardization—the notion that globalization homogenizes standards of all kinds, such as business cultures or built environments (no matter how good the architecture).

The economic trajectory and switching illustrated by the case of Chicago contests the thesis of the homogenizing effects of today's advanced economic sectors, a thesis which also brings with it an emphasis on intercity and interregional competition. This thesis and its implications could also be extended to certain types of regions and megaregions with similarly specialized economic trajectories, albeit with very different contents. The Chicago case shows that becoming part of a knowledge economy is not simply a question of dropping a manufacturing and agro-industrial past, and then proceeding to converge on the headquarters-services-cultural sector axis. Just as critical is executing the switch described

earlier—whatever might be the specifics of an area's past.[24]

Further, Chicago also indicates that the meaning of homogenized urban and regional landscapes needs to be examined empirically. It becomes critical to establish the particular specialized sectors that might inhabit that homogenized landscape.

HOMOGENIZED BUILT ENVIRONMENTS:OBSCURING ECONOMIC DIFFERENCES

Here I will argue that to some degree these homogenized and convergent state-of-the art urban and increasingly regional landscapes are actually functioning as an "infrastructure." As an infrastructure, these homogenized built environments guarantee the provision of all advanced systems and luxuries needed/ desired by the firms and households in leading economic sectors are in place. Office districts, high-end housing and commercial districts, conventional and digital connectivity, cultural districts, security systems, airports, and so on, are all in place and they are all state-of-the-art.

Comparative analyses rely on similarities and differences to make their point. Contemporary urbanization, whether at the urban, metro, or regional level, is often seen as marked by a homogenizing of the urban landscape and a growing range of its built environments. This is especially so in the case of global cities and global city-regions due to the intensity and rapidity of urban reconstruction in such areas. And yet this obscures the fact of the diversity of economic trajectories through which cities and regions emerge and develop (as discussed in the preceding section), even when the final visual outcomes may look similar. Out of this surface analysis based on homogenized landscapes and built environments, comes a second possibly spurious inference, that this homogenizing is a function of economic convergence, for instance, the notion that we are all moving to (the same) knowledge economy. Both propositions—that similar visual landscapes are indicators of both similar economic dynamics and of convergence—may indeed capture various situations. But these propositions also obscure key conditions that point to divergence and specialized differences; in fact, divergence and specialized difference are easily rendered invisible by such notions. We need to take such spurious inferences into account when understanding the character of these megaregions.

At the most general level we might start with developments at the macro-economic level that can easily lead observers to buy into the homogenization thesis. An important structural trend evident in all reasonably working economies is the growing service intensity in the organization of just about all economic sectors, including rather routine and often nonglobalized sectors. Whether a firm is in mining and agriculture, manufacturing, or a service industry such as transport or health, more of these firms are buying more producer services. Some of this translates into a growing demand for producer services in global cities, but much of it translates into a demand for such services from regional centers, albeit often less complex and advanced versions of those services.

The growth in the demand for producer services is a structural feature of advanced market economies that affects most economic sectors. It is not just a feature of globalized sectors.[25] What globalization brings to this trend is a sharp increase in the demand for complexity and diversity of professional knowledge.[26] It is this qualitative difference that leads to the heightened agglomeration economies evinced by firms in global cities compared to other types of urban areas. But the basic structural trend is present in both types of areas. This perspective also clarifies what is a somewhat misguided interpretation about the higher growth rates of producer services in cities that are not global. The trend is to assume ipso facto that these higher growth rates of producer services reflect decline and/or the departure of producer services from global cities. Those higher growth rates are actually in good part the result of lagged growth of these services throughout the national economy; global cities had their extremely high growth rates much earlier, in the 1980s.[27] The lower growth rates evident in global cities compared with other cities should thus not necessarily be interpreted as losses for the former, but rather as the latter entering this new structural phase of market economies.[28] Looking at matters this way recodes some common interpretations of growth and decline.

What is critical for the analysis in this section is that the growth of this intermediate economy across diverse urban areas amounts to a kind of structural convergence that explains a homogenizing of built environments and spatial patterns even when the sectors serviced are radically different. Regardless of economic sector and geographic location, firms are buying more of these services. A mining firm, a transport firm, and a software firm all need to buy legal and accounting services. To some extent these services may be produced in the same city and in similar built environments, even though they are feeding very different economic sectors and geographic sites of the larger economy, including the megaregional economy. Thus "old economy" sectors such as manufacturing and mining are also feeding the growth of the intermediate economy.

This structural convergence does filter through and homogenizes spatial organization and the visual order of the built environment. It does account for key patterns evident in cities small and large, notably the well-documented growth of a new type of professional class of young urbanites and the associated high-income gentrification and growth of the cultural sector. This convergence and homogenizing of the visual order easily obscures the specific trajectories and contents through which a region develops a knowledge economy, as discussed in the preceding section of this paper.

Seen this way, we can begin to qualify the homogenization and convergence thesis. There is a kind of convergence at an abstract systemic level, and at the level of the needed built environments for the new intermediate economy and the new kinds of professional workforces. But at the concrete, material interface of the economy and its built environments, the actual content of the specialized services that inhabit that built environment can vary sharply.

From here, then, my proposition that critical components of the homogenized/convergent urban and regional landscape frequently presented as today's quintessential new advanced built environment, are actually more akin to an infrastructure for economic sectors. This unsettles the concept (and the reality) of the built environment as we have generally used it. The critical question becomes what inhabits that "infrastructure?" Looking similar does not necessarily entail similar contents, circuits, moments of a process. This illustrates the thesis that different dynamics can run through similar institutional and spatial forms, and vice versa.[29] Thus the substantive character of convergence in the global city model, for instance, is not the visual landscape per se but its function as an infrastructure; and it is, above all, the development and partial importation of a set of specialized functions and the direct and indirect effects this may have on the larger city, including its built environment.

One question here is whether this distinction between homogenized built environments and the often highly diverse contents they house also need to become part of our understanding of what is specific to a megaregion?[30] State-of-the-art office buildings or speed rail or airports can look very similar yet serve very different economic sectors. These types of differences are becoming increasingly important to understand a city's, a region's, and possibly a megaregion's place in the global economy. There are two reasons for this. One is the shift from a Keynesian spatial economy striving for national territorial convergence to a post-Keynesian space economy oriented towards territorial targeting (global cities, silicon valleys, science parks, and so on). The second is that a city's, a region's, and possibly a megaregion's advantage in the global economy is a function of positioning in multiple highly particularized, and often very specialized, economic circuits; it is not helpful to think of "the" place of "the" megaregion in "the" global economy.

SCALING AND ITS CONSEQUENCES

Moving from the scale of the city to that of an urbanized region alters the analytics. A region easily contains sites that evince agglomeration economies and sites that offer the option of geographic dispersal of activities. Beyond this, questions of power and inequality play out rather differently when we compare regions and cities. To sharpen the focus, I confine the discussion in this section of the paper to global cities and global-city regions. The concept of the global-city region adds a whole new dimension to questions of territory and globalization.[31] This type of comparison illustrates some of the issues developed in the more analytic discussion of the preceding sections. And it makes the argument in a more descriptive manner, so even if a reader rejects the analytics of the preceding section it still leaves room for the empirics.

A first difference concerns the question of territory. The territorial scale of the region is far more likely to include a cross-section of a country's economic activities than the scale of the city. For instance, it is likely to include as key variables manufacturing and a range of standardized economic sectors that are at the heart of the national economy. This, in turn,

brings with it a more benign manifestation of globalization. The concept of the global city introduces a far stronger emphasis on strategic components of the global economy, typically subject to extreme agglomeration economies in top-level management functions and specialized corporate servicing; this in turn can lead to extreme forms of power and inequality in the global city. Secondly, the concept of the global city will tend to have a stronger emphasis on the networked economy because of the nature of the industries that tend to be located there: finance, media, and other specialized services. And, thirdly, it will tend to have more of an emphasis on economic and spatial polarization because of the disproportionate demand for very high- and very low-income jobs in these cities compared with what would be the case for the region which would have far more middle-range firms and workers.

Overall, I would say, the concept of the global city is more attuned to questions of power and inequality. The concept of the global-city region is more attuned to questions about the nature and specifics of broad urbanization patterns, a more encompassing economic base, more middle sectors of both households and firms, and hence to the possibility of having a more even distribution of economic benefits under current economic growth dynamics, including economic globalization. In this regard, it could be said that the concept of the global-city region allows us to see the possibilities for a more distributed kind of growth, a wider spread of the benefits associated with economic growth, including growth resulting from globalization.

Secondly, both concepts have a problem with boundaries of at least two sorts: the boundary of the territorial scale as such and the boundary of the spread of globalization in the organizational structure of industries, institutional orders, places, and so on. In the case of the global city I have opted for an analytic strategy that emphasizes core dynamics rather than the unit of the city as a container—a container being an entity that requires territorial boundary specification. Emphasizing core dynamics and their spatialization (in both actual and digital space) does not completely solve the boundary problem, but it does allow for a fairly clear trade-off between emphasizing the core or center of these dynamics and their spread institutionally and spatially. In my work I have sought to deal with both sides of this trade-off: by emphasizing, on the one side of the trade-off, the most advanced and globalized industries, such as finance, and, on the other side, how the informal economy (typically seen as local) in major global cities is articulated with some of the leading industries. In the case of the global-city region, it is not clear to me how Scott (2001) specifies the boundary question both in its territorial sense and in terms of its organization and spread.

A third difference is the emphasis on competition and competitiveness, much stronger in the global-city region construct. In my reading, the nature itself of the leading industries in global cities strengthens the importance of cross-border networks and specialized divisions of functions among cities in different countries and/or regions rather than international competition per se. Further, though competitiveness is a necessary condition, it is far less prominent in these sectors which tend to flag "talent" as their key rather than competence, than it would be in developing regional rapid rail where the question of competence (rather than speculative talent, for instance) is essential. Global finance and the leading specialized services catering to global firms and markets—law, accounting, credit rating, telecommunications—constitute cross-border circuits embedded in networks of cities, each possibly part of a different country. It is a de-facto global system, centered in more than competition and competitiveness.

The industries that will tend to dominate global-city regions are less likely to be networked along these lines. For instance, in the case of large manufacturing complexes, and of final and intermediate consumption complexes, the identification with the national is stronger and the often stronger orientation to consumer markets brings to the fore the question of quality, prices, and the possibility of substitution. Hence competition and competitiveness are likely to be far more prominent. Further, even when there is significant offshoring of production and in this regard an international division of production, as in the auto industry, this type of internationalization tends to be in the form of the chain of production internal to a given firm, which today can cross borders. Insofar as most firms still have their central headquarters associated with a specific region and country, the competition question is likely to be prominent and, very importantly, sited—i.e., it is the U.S. versus the Japanese auto manufacturers, though even this is changing.

The question of the competitiveness of a region is deeply centered in its conventional infrastructure—transport of all sorts, water and electricity supply and distribution, airports, and so on. To some extent this is also a crucial variable in the case of global cities, but it is a far more specialized type of infrastructure in the latter. The regional scale brings to the fore questions of public transport, highway construction, and kindred aspects in a way that the focus on global cities does not. Again, it reveals to what extent a focus on the region produces a more benevolent representation of the impacts of the global economy. A focus on the regional infrastructure is far more likely to include strong consideration of middle-class needs. In contrast, a focus on the global city will tend to bring to the fore the growing inequalities between highly provisioned and profoundly disadvantaged sectors and spaces of the city, and hence questions of power and inequality.

A fourth difference, connected to the pre-ceding one, is that a focus on networked cross-border dynamics among global cities also allows us to capture more readily the growing intensity of such transactions in other domains—political, cultural, social, criminal.[32] We now have evidence of greater cross-border transactions among immigrant communities and communities of origin and a greater intensity in the use of these networks once they become established, including for economic activities that had been unlikely until now. We also have evidence of greater cross-border networks for cultural purposes, as in the growth of international markets for art and a transnational class of curators; and for

nonformal political purposes, as in the growth of transnational networks of activists around environmental causes, human rights, and so on. These are largely city-to-city cross-border networks, or, at least, it appears at this time to be simpler to capture the existence and modalities of these networks at the city level. The same can be said for the new cross-border criminal networks. Dealing with the regional scale does not necessarily facilitate recognizing the existence of such networks from one region to the other. It is far more likely to be from one specific community in a region to another specific community in another region, thereby neutralizing the mean-ing of the region as such.

One key implication of this comparison is that we need to control for some of the inevi-table differences that are a function of scale per se. There is a risk of reifying the spatial organization of a bounded terrain, such as the megaregions identified by RPA. Compar-ing a city and a region does add important information to our effort of understanding the variability of location and of the advantages of proximity. But it is also a fact that the reality of a megaregion may well rest on dynamics that underlie both of these—city and region.

Part of the task of specifying megaregions needs to get at these sharp differences within a region and at the possibly shared dynamics underlying these differences: thus the multipolarity and geographic dispersal that characterize these megaregions may in part also feed agglomeration economies in these regions' cities arising precisely out of that dispersal. A critical question is whether some of these diverse formations—multipolarity, dispersal, and agglomeration—can be re-regionalized. This can take two forms, one more elementary and one more complex. The elementary one is increasing the range of formations that could be incorporated within a megaregion, rather than only thinking in terms of the high-end of an economic sector or a firm's operations, as is often done. The complex form is to increase the range of formations that are part of a given growth sector or a firm's multi-sited chain of operations. Such a re-regionalizing of the components of economic growth could emerge as a major advantage of megaregions.

Notes

1. For one of the definitive examinations of the shortcomings of the data on sub-national scalings see the report by the National Academy of Sciences (2003). See also generally, OECD 2006; 2007.

2. The 10 megaregions include the Arizona Sun Corridor, Cascadia (Pacific Northwest), Florida, the Great Lakes, the Gulf Coast, the Northeast, Northern California, Piedmont Atlantic (Southeast U.S.), Southern California, and the Texas Triangle.

3. Besides "regionalizing" various segments of a firm's chain of operations, one might also propose to regionalize more segments of various commodity chains. See, for instance, Gereffi, Gary, John Humphrey, and Timothy Sturgeon. 2005. "The Governance of Global Value Chains." *Review of International Political Economy* (Special Issue: Aspects of Globalization). 12 (1): 78–104.

4. For one of the best data sets on the dispersal at the global scale of the operations of firms in corporate services see Globalization and World Cities Study Group and Network (GAWC.) 1998. http://www.lboro.ac.uk/departments/gy/research/gawc.html.

5. Regional Plan Association (2006). "America 2050: A Prospectus." New York.

6. A parallel issue here, not fully addressed in this paper, is the articulation of technical connectivity with social connectivity. See, for instance, Garcia, D. Linda. 2002. "The Architecture of Global Networking Technologies." In *Global Networks/Linked Cities*, edited by Saskia Sassen. New York and London: Routledge, p. 39–69.

7. This is a type of agglomeration economy I found in my research on global cities, but it can also be applied to national or regional scales. The hypothesis was that the greater the capabilities for geographic dispersal a firm has, the higher the agglomeration economies it is subject to in some of its components, notably top-level headquarter functions.(See *The Global City*. Princeton: Princeton University, 2001, 2nd ed.; original edition 1991: new preface) for a brief explanation of the nine hypotheses that specify the global city model. It is the most specialized functions pertaining to the most globalized firms which are subject to the highest agglomeration economies. The complexity of the functions that need to be produced, the uncertainty of the markets such firms are involved in, and the growing importance of speed in all these transactions, is a mix of conditions that constitutes a new logic for agglomeration; it is not the logic posited in older models, where weight and distance (cost of transport) are seen to shape agglomeration economies. The mix of firms, talents, and expertise in a broad range of specialized fields, makes a certain type of dense environment function as a strategic knowledge economy wherein the whole is more than the sum of (even its finest) parts.

8. The region, the metro area and the city are scalings that allow us to capture the many highly specialized circuits that are comprised by what we call "the" global economy. Different circuits contain different groupings of regions and cities. Viewed this way, the global economy becomes concrete and specific, with a well-defined geography. Goods and services are redistributed to a vast number of destinations, no matter how few the points of

origin are in some cases. With globalization, this capacity to redistribute globally has grown sharply). By focusing on a scale such as the region and on the diverse types of economic spaces it contains, we can capture multiple points of redistribution, as well as points of origin. For a definitive treatment of some of these issues as they apply to service industries see Peter J. Taylor. 2003. *World City Network: A Global Urban Analysis.* New York: Routledge.

9. For an analysis of options see Henderson, Jeffrey. 2005. "Governing growth and inequality: the continuing relevance of strategic economic planning." Pp. 227–36 in *Towards a Critical Globalization Studies,* edited by R. Appelbaum and W. Robinson. New York: Routledge.

10. But there is a caveat. A second key hypothesis I developed to specify the global city model is that the more headquarters actually buy some of their corporate functions from the specialized services sector rather than producing them in house, the greater their locational options become. Among these options is moving out of global cities, and more generically, out of dense urban environments. This is an option precisely because of the existence of a networked specialized producer services sector that can increasingly handle some of the most complex global operations of firms and markets. It is precisely this specialized capability to handle the global operations of firms and markets that distinguishes the global city production function in my analysis, not the number per se of corporate headquarters of the biggest firms in the world, as is often suggested.

11. One of the best and most detailed analysis comparing two different formats for high-tech districts is Saxenian, Anna-lee. 1996. *Regional Advantage: Culture and Competition in Silicon Valley and Route 128.* Cambridge, MA: Harvard University Press.

12. This also allows us to go beyond many of the tropes in this subject. For instance, it contests the key proposition of the L.A. model on urban form and the new economy: that dispersal is the spatial form of advanced sectors. The facts on the ground (including for the L.A. region) show both dispersal and spatial concentrations. For an in-depth analysis of these issues see Michael Conzen and Richard Greene (eds) "The L.A. and Chicago Schools: A debate." Special Issue of *Progressive Geography* (2007). The available evidence, and there is plenty

of it, indicates that key factors shaping the spatial organization of leading firms are present in both the L.A. region and in older cities such as Chicago and New York. But the larger spatial organization of each region gets coded differently. What is coded as multipolarity in the L.A. region gets coded as "relocation to the metropolitan area or beyond" in Chicago and New York. At the same time, dense concentrations of the most innovative and globalized sectors subject to agglomeration economies are present in both L.A. and Chicago/New York, but their contents are very different, a subject I return to later.

13. And, indeed, certain very contemporary forms of dispersal are a function of particular capacities developed in settings marked by high agglomeration economies (exemplified by global cities). And they are not only happening in the narrowly understood sphere of the economy: thus we see the growth of an international curatorial class, and major museums allowing their most valued collections to go on tour in a foreign country; this is enabled by the development of a specialized servicing capability (law, accounting, insurance, finance)—the global city economic production function.

14. This spatial lens is also to be distinguished from the more common angle of firms and markets (see, for example, Dieter Ernst. 2005. "The New Mobility of Knowledge: Digital Information Systems and Global Flagship Networks." Pp. 89–114 in *Digital Formations: IT and New Architectures in the Global Realm,* edited by Robert Latham and Saskia Sassen. Princeton: Princeton University Press).

15. For a full development of this subject please see Sassen 2006a (*Territory, Authority, Rights: From Medieval to Global Assemblages.* Princeton, NJ: Princeton University Press): chs. 5 and 7; 2001 (op. cit.).

16. Elsewhere I have explained in detail this thwarting of technical logics by the economic, financial, or for that matter cultural and political logics of users; see Sassen 2006a (op.cit): ch 7.

17. Today's multinationals have over one million affiliates worldwide. Affiliates are but one mode of global operation. For empirical details about the range of formats of global operations see Sassen 2006b. *Cities in a World Economy* 3rd.ed. (Sage/PineForge): ch.2; Taylor, Peter J. 2004. *World City Network: A Global Urban Analysis.* New York: Routledge; World Federation of Exchanges, 2007.

"Annual Statistics for 2006." Paris: World Federation of Exchanges, (and annual updates).

18. For a detailed examination of the importance of the subnational scale for a global market, see Harvey, Rachel. 2007. "The Sub-National Constitution of Global Markets." In *Deciphering the Global: Its Spaces, Scales and Subjects.* Edited by S. Sassen. New York and London: Routledge.

19. See, for instance Rutherford, Jonathan, 2004. *A Tale of Two Global Cities: Comparing the Territorialities of Telecommunications Developments in Paris and London.* Aldershot, UK and Burlington, VT: Ashgate.

20. For a full development of these patterns see Sassen 2006b (op.cit.): ch.5.

21. For one of the most detailed examinations of the current and past economic patterns of Chicago and its region see Greene (ed) 2006.

22. This brings to the fore the specialized division of functions in the global economy, one partly constituted and implemented through a proliferation of specialized cross-border city-networks. The critical mass of these networks has expanded to include about 40 major and minor global cities. There are many networks and different types of functions/positions for cities. Detecting this has required developing new methodologies (see Taylor 2004 (op. cit.); Alderson, Arthur S. and Jason Beckfield. "Power and Position in the World City System." *American Journal of Sociology* 109(4): 811–51; and the illuminating debate on questions of method between Taylor and Alderson/Beckfield, forthcoming in *The American Journal of Sociology*). The global network of cities is much more than just a set of cross-border flows connecting cities. It is a complex, highly specialized organizational infrastructure for the management and servicing of the leading economic sectors.

23. In Sassen (2006a) I develop this notion of switching (existing capabilities switching to novel organizing logics) in order to understand the formation of today's global economy as well as today's partial denationalizing of state capacities.

24. For very different types of cities and economic trajectories, see for example Amen, Mark M., Kevin Archer, and M. Martin Bosman (eds), 2006. *Relocating Global Cities: From the Center to the Margins.* New York: Rowman & Littlefield; and

Gugler, Joseph. 2004. *World Cities Beyond the West.* Cambridge: Cambridge University Press.

25. For one particular aspect—artistic practice as it feeds into commercialized design—see Lloyd, Richard. 2005. *NeoBohemia: Art and Bohemia in the Postindustrial City.* London and New York: Routledge.

26. In developing the global city model I posited that a critical indicator is the presence of a networked, specialized producer services sector capable of handling the global operations of firms and markets, whether national or foreign. Given measurement difficulties, a proxy for this networked sector is the incidence and mix of producer services in a city. This is frequently reduced to the share of producer services employment as the indicator of global city status. This is fine, though it needs empirical specification as to the quality and mix of the producer services industries. More problematic is to interpret a small share, or a declining share, or a falling growth rate, or a lower growth rate than in non-global cities, as an indicator of global city status decline or as signaling that the city in question is not a global city. Similarly problematic is a variant on this indicator is the share a city has of national employment in producer services and whether it has grown or fallen; the notion here is that if a city such as New York or London loses share of national employment in producer services, it loses power.

27. On that earlier phase, see, for example Drennan, Mathew P. 1992. "Gateway Cities: The Metropolitan Sources of U.S. Producer Service Exports." *Urban Studies* 29(2):217–35.

28. Thus the high growth rates of producer services in smaller cities as compared with global cities is not necessarily a function of relocations from global cities to better-priced locations. It is a function of the growing demand by firms in all sectors for producer services. When these services are for global firms and markets, their complexity is such that global cities are the best production sites. But when the demand is for fairly routine producer services, cities at various levels of the urban system can be adequate production sites. The current spatial organization of the producer services reflects this spreading demand across economic sectors.

29. In Sassen (2006a) I posit a parallel argument for the liberal state as it is subjected to the forces of economic and political globalization. The outcome

does not necessarily mean that these states lose their distinctiveness, but rather that they implement the necessary governance structures to accommodate global projects and that they do so through the specifics of their state organization.

30. For a detailed examination of this mix of visual, urban engineering, architectural, and economic issues across 16 major cities in the world see Burdett, Ricky (ed). 2006. *Cities: People, Society, Architecture.* New York: Rizzoli; Sudjic, Deyan. 1993. *The Hundred Mile City.* Harvest/HBJ; and 2005. *The Edifice Complex: How the Rich and Powerful Shape the World.*

31. For a development of this concept see Scott, A. J. 2001. *Global City-Regions.* Oxford: Oxford University Press.

32. I cannot resist referring to a book that breaks new terrain in this regard: Hagedorn, John, ed. 2004. *Gangs in the Global City: Exploring Alternatives to Traditional Criminology.* Chicago: University of Illinois at Chicago.

Appendix A:
Discussion Summaries

PANELIST AND DISCUSSION COMMENTS

Robert Yaro, the president of the Regional Plan Association (RPA), which cosponsored the conference along with Princeton's Policy Research Institute for the Region, introduced the day's discussion with an overview of the concept of megaregions. He said RPA and other organizations and universities are engaged in a project of 21st-century planning for America's megaregions. One of those efforts is called the "National Committee for America 2050." Yaro said of the committee that "the goal is to develop a framework for a national growth strategy organized around megaregions."

He noted that there is precedent for such planning in parts of Europe and Asia, which have successfully charted economic zones. Europeans have established "global integration zones"; Asian cities and countries have constructed complex rail networks, creating corridors that promote economic development and housing development, Yaro said.

Yaro asserted that megaregions already exist in the United States; it is up to organizations and local and regional governments to take advantage of them for economic gain. If state and federal authorities do not tackle challenges that span across multiple cities and towns

—transportation, housing, and economic growth—the U.S. will have difficulty competing globally.

He explained some of the current national population trends: There are several areas that are projected to have population growth of more than 100 percent between 2000 to 2050; the South and the West are growing robustly; other places will "hollow out" and experience population losses of up to 15 percent, such as the area known as the Rust Belt, parts of the Northeast, and North Central industrial regions.

Accurate assessment of population trends is crucial to the type of analysis necessary for policy recommendations and for estimates concerning certain aspects of the economy.

Commenting on the papers and the notion of megaregions overall, **Kip Bergstrom**, executive director of the Rhode Island Economic Policy Council, stressed the necessity of strong transportation policy. In particular, he is focused on regional rail lines that can connect various parts of megaregions to one another, short of national coordination with Amtrak-type scope.

With a nod toward Yaro and **Edward Glaeser**'s mention of the role of cars in 20th-century agglomeration, Bergstrom noted that cars can't go any faster than they already do, so today's megaregions have to be facilitated with rail lines. Commuter rail needs to move at Acela-type speeds but hopefully with better technology, he said, referring to Amtrak's beleaguered high-speed line. These regional lines could and should span several towns and even states, boosting economic growth as laborers and consumers reach destinations easier and faster. He noted that such rail lines help with planned building and landscape development because fewer interruptive roads would be built.

Bergstrom also challenged the audience to think about the existence of megaregions in a world with a handful of other simultaneous trends, leading him to question to what degree past trends of population growth and density are really a prelude to the future. He listed climate change, globalization, and emerging economic changes as having dynamic effects with the issues related to megaregions. Is the growth in some of the recent high-growth areas Glaeser recited sustainable if society gets serious about restricting carbon emissions, Bergstrom asked?

Professor of economics and international affairs and *New York Times* columnist **Paul Krugman** joined the discussion by joking that the megaregion trend has demonstrated that New York and Philadelphia are indeed the two primary cities of New Jersey.

He sought to apply economic analysis to the question of whether "megaregions" qualify as distinct economic entities.

He said that one classic definition of an economic unit is one in which there are linkages, labor-market mobility, and knowledge spillover. He said there are linkages in what would be described as megaregions, but they are much more diffuse than one would anticipate for a real metropolitan area. Is commuting between Pennington, New Jersey, and Bucks County, Pennsylvania, and vice versa, for instance, representative of true linkages or is it just cross-commuting between the catchment areas? He also said that megaregions do not necessarily give rise to the externalities normally anticipated from a singular economic entity.

However, Krugman asserted, the massive sprawl does create the need for public goods —one criterion for existence as an economic unit. As a result of the public-good necessity of these regions—particularly in transportation —planning, policymaking, and economic analysis could be crucial and indeed have an effect on megaregions' future.

He said he agrees with **Saskia Sassen** that the outsourcing of back-office functions is occurring with regularity but that it is crucial to have people near one another for other jobs. He rounded out his remarks by wondering—with a disclaimer that his view might reflect "Northeast chauvinism"—whether the megaregions posited to exist in other areas of the country really have the same dynamics as the Northeast corridor.

A roundtable discussion after the paper presentations and the responses produced provocative audience input, with experts in business and urban planning participating in a dialogue about the opportunities and challenges from the development of megaregions.

Most notable was the participation of **Robert Geddes**, the former—and first—dean of Princeton University's School of Architecture, who pressed the panel members to further explore job creation as the crux of the issue around which megaregions and planning revolve.

Almost everyone in the discussion, which Yaro moderated, agreed that governmental and nongovernmental institutions ought to recognize the importance of sprawl "megaregions"—with whatever labels and designations we choose—and start engaging in semi-regional, regional, and sometimes national planning for the attendant economic, land development, and transportation issues.

One audience member, however, added a note of caution, challenging everyone to consider that organizations, companies, and governments might not currently have the capacity, resources, and desire to take on such sweeping initiatives.

Appendix B:
Conference Agenda

The Economic Geography of Megaregions

FEBRUARY 9, 2007

Sponsored by the Policy Research Institute for the Region at Princeton University and the Regional Plan Association

Welcome
Nathan B. Scovronick, Acting Director, Policy Research Institute for the Region

Emerging American Megaregions and Public Policy Implications
Robert D. Yaro, President, Regional Plan Association

Do Regional Economies Need Regional Coordination?
Edward Glaeser, Fred and Eleanor Glimp Professor of Economics, Department of Economics, Harvard University

The Economic Geography of Global Megacities and Megaregions
Saskia Sassen, Ralph Lewis Professor of Sociology, University of Chicago

Responses to Research Papers
Kip Bergstrom, Executive Director, Rhode Island Economic Policy Council
Paul R. Krugman, Professor of Economics and Public Affairs, Princeton University

Closing Remarks
Robert D. Yaro, President, Regional Plan Association

Appendix C:
Participant Biographies

Kip Bergstrom
Executive Director, Rhode Island Economic Policy Council

Kip Bergstrom has served since May 1998 as executive director of the Rhode Island Economic Policy Council. Prior to that, he was economic development director in Stamford, Connecticut, from 1993 to 1997, where he helped to reduce Stamford's office and industrial vacancy rates by 80 percent and closed 45 deals representing over 6,000 jobs, including the 2,200-employee North American headquarters of UBS Warburg Dillion Read, the investment banking arm of the world's second-largest bank. Bergstrom has a master's in city and regional planning from the John F. Kennedy School of Government at Harvard University, where he was the first student to specialize in economic development.

Edward Glaeser
Fred and Eleanor Glimp Professor of Economics, Harvard University

Edward Glaeser is the Fred and Eleanor Glimp Professor of Economics in the Faculty of Arts and Sciences at Harvard University, where he has taught since 1992. He is director of the Taubman Center for State and Local Government and director of the Rappaport Institute of Greater Boston. He teaches urban and social economics and microeconomic theory. He has published dozens of papers on cities, economic growth, and law and economics. In particular, his work has focused on the determinants of city growth and the role of cities as centers of idea transmission. He also edits the *Quarterly Journal of Economics.* He received his Ph.D. from the University of Chicago in 1992.

Keith S. Goldfeld
Program Manager, Policy Research Institute for the Region, Princeton University

Before joining Princeton University's Policy Research Institute for the Region as program director, Keith S. Goldfeld worked as a policy and data analyst for a number of organizations spanning the government, not-for-profit, and private sectors. Before coming to the institute, he was a health care consultant at Gold Health Strategies, Inc. in New York City. His experience in government includes working as a policy analyst in various departments at the Port Authority of New York and New Jersey, as a senior budget and policy analyst at the New York City Independent Budget Office, and as a consultant for the New York City Board of Education. While at HealthFirst, a not-for-profit, hospital-owned managed-care organization in New York, he served as the organization's director of analysis. Goldfeld received his B.A. in computer science from Williams College, his master's of public affairs and urban and regional planning from the Woodrow Wilson School at Princeton, and his M.S. in statistics from Baruch College/City University of New York.

Paul R. Krugman
Professor of Economics and International Affairs, Princeton University

The author or editor of dozens of books and several hundred articles, primarily about international trade and international finance, Paul R. Krugman is also nationally known for his twice-weekly columns in *The New York Times* and his monthly columns in *Fortune Magazine* and *Slate*. He was the Ford International Professor of International Economics at the Massachusetts Institute of Technology and has served on the U.S. Council of Economic Advisers. He was the recipient of the 1991 John Bates Clark Medal, an award given every two years by the American Economic Association to an economist under 40. He received his Ph.D. from the Massachusetts Institute of Technology.

Saskia Sassen
Ralph Lewis Professor of Sociology, University of Chicago; Centennial Visiting Professor, London School of Economics

Saskia Sassen is the Ralph Lewis Professor of Sociology at the University of Chicago, and Centennial Visiting Professor at the London School of Economics. Her new book is *Territory, Authority, Rights: From Medieval to Global Assemblages* (Princeton University Press 2006). She has just completed for UNESCO a five-year project on sustainable human settlement for which she set up a network of researchers and activists in over 30 countries; it is published as one of the volumes of the *Encyclopedia of Life Support Systems* (Oxford, U.K.: EOLSS Publishers) [http://www.eolss.net]. Other recent books are the third fully updated *Cities in a World Economy* (Sage 2006), *A Sociology of Globalization* (Norton 2007), and the co-edited *Digital Formations: New Architectures for Global Order* (Princeton University Press 2005). *The Global City* came out in a new fully updated edition in 2001. Her books are translated into 16 languages. Sassen serves on several editorial boards and is an adviser to several international bodies. She is a member of the Council on Foreign Relations, a member of the National Academy of Sciences Panel on Cities, and chair of the Information Technology and International Cooperation Committee of the Social Science Research Council (USA). Her comments have appeared in *The Guardian, The New York Times, Le Monde Diplomatique*, the *International Herald Tribune, Newsweek International, Vanguardia, Clarin*, the *Financial Times*, among others.

Nathan B. Scovronick
Acting Director, Policy Research Institute for the Region; Director of the Undergraduate Program and Lecturer of Public and International Affairs, Woodrow Wilson School

Nathan B. Scovronick is the acting director of the Policy Research Institute for the Region and the director of the Undergraduate Program at the Woodrow Wilson School of Public and International Affairs at Princeton University. He was the director of the Program in New Jersey Affairs at the Woodrow Wilson School from 1993 to 1995. He previously served as executive director of the Treasury Department of the State of New Jersey and as deputy director of the New Jersey General Assembly. He is the coauthor, (with Jennifer L. Hochschild) of *The American Dream and the Public Schools* (Oxford University Press, 2003).

Petra Todorovich
Director, America 2050

Petra Todorovich is director of America 2050, a national program based at Regional Plan Association. This new initiative addresses the anticipated 40 percent increase in population growth in the U.S. by the year 2050 by strengthening emerging megaregions—extended networks of metropolitan areas—where most of the population and employment growth will be focused in the coming decades.

Prior to launching America 2050, Todorovich directed RPA's Region's Core program and coordinated the "Civic Alliance to Rebuild Downtown New York," a network of organizations that came together shortly after 9/11 to promote the rebuilding of the World Trade Center site and Lower Manhattan. She has planned numerous public forums and workshops for the rebuilding of the World Trade Center and Lower Manhattan, including the 2002 "Listening to the City" meetings at the Javits Center that brought more than 4,500 people together to consider plans for the World Trade Center site. She authored the 2004 "Civic Assessment of the Lower Manhattan Planning Process" and other pieces of analysis on the rebuilding process and New York City development. Todorovich received her B.A. from Vassar College, where she studied geography and French, and a master's in city and regional planning from the Bloustein School of Planning and Public Policy at Rutgers University.

Robert D. Yaro
President, Regional Plan Association

Robert D. Yaro is the president of Regional Plan Association, where he has been on the staff since 1990. Headquartered in Manhattan and founded in 1922, RPA is America's oldest and most respected independent metropolitan research and advocacy group. Yaro is also the Practice Professor in City and Regional Planning at the University of Pennsylvania. Formerly he served on the faculties of Harvard

University and the University of Massachusetts—Amherst. He chairs *The Civic Alliance to Rebuild Downtown New York*, a broad-based coalition of civic groups formed to guide redevelopment in Lower Manhattan in the aftermath of the September 11 attacks on the World Trade Center. He is also a director of Alliance for Regional Stewardship. Yaro is an honorary member of the Royal Town Planning Institute. He holds a master's degree in city and regional planning from Harvard University and a B.A. in urban studies from Wesleyan University.